Gardner Dexter Hiscox

Gas, Gasoline, and Oil Vapor Engines

A New Book Descriptive of Their Theory and Power....

Gardner Dexter Hiscox

Gas, Gasoline, and Oil Vapor Engines
A New Book Descriptive of Their Theory and Power....

ISBN/EAN: 9783337025489

Printed in Europe, USA, Canada, Australia, Japan

Cover: Foto ©berggeist007 / pixelio.de

More available books at **www.hansebooks.com**

GAS, GASOLINE

AND

OIL VAPOUR ENGINES

*A NEW BOOK DESCRIPTIVE OF THEIR THEORY
AND POWER, ILLUSTRATING THEIR DESIGN,
CONSTRUCTION AND OPERATION*

FOR

STATIONARY, MARINE AND VEHICLE
MOTIVE POWER

BY

GARDNER D. HISCOX, M.E.

A WORK DESIGNED FOR THE GENERAL INFORMATION OF EVERY ONE INTERESTED
IN THIS NEW AND POPULAR PRIME-MOVER, AND ITS ADAPTATION TO
THE INCREASING DEMAND FOR A CHEAP, SAFE AND
EASILY MANAGED MOTIVE POWER

London:
E. & F. N. SPON, Limited, 125 STRAND
New York:
NORMAN W. HENLEY & CO., 132 NASSAU STREET
1897

PREFACE.

THE entire absence of any literature on Explosive Motors as made in the United States, other than such as has been published from time to time in journals and magazines, and in view of the constant inquiry of persons interested in the use of motive power as to the various styles and designs of gas, gasoline and oil engines and the principles of their working, have induced the author of this work to put into a practical shape for the ordinary reader the principles and practice of the operation of this class of motors as made principally in the United States. The German, French and English books and their reprints in this country scarcely allude to American practice or American engines.

The author has been greatly favoured by a large number of explosive-motor builders in the United States for illustrations and details of the motors of their manufacture and their operation. He hopes by the publication of this work that many inquiries will be answered, and that seekers for small power will find in the explosive motor the economical prime-mover so much desired.

GARDNER D. HISCOX.

NEW YORK: *January* 1, 1897.

CONTENTS.

CHAPTER		PAGE
I.	INTRODUCTORY	1
	HISTORICAL	3
II.	THEORY OF THE GAS AND GASOLINE ENGINE	7
III.	UTILISATION OF HEAT AND EFFICIENCY IN GAS ENGINES	18
IV.	RETARDED COMBUSTION AND WALL-COOLING	25
V.	CAUSES OF LOSS AND INEFFICIENCY IN EXPLOSIVE MOTORS	33
VI.	ECONOMY OF THE GAS ENGINE FOR ELECTRIC LIGHTING	37
	THE MATERIAL OF POWER IN EXPLOSIVE ENGINES	41
VII.	PETROLEUM PRODUCTS USED IN EXPLOSIVE ENGINES	45
	CARBURETTERS	47
VIII.	CYLINDER CAPACITY OF GAS AND GASOLINE ENGINES	54
	MUFFLERS	56
IX.	GOVERNORS	58
X.	IGNITERS AND EXPLODERS	65
XI.	CYLINDER LUBRICATION	81
XII.	ON THE MANAGEMENT OF EXPLOSIVE MOTORS	84
	THE MEASUREMENT OF POWER	89
XIII.	THE INDICATOR AND ITS WORK	96
	VIBRATION OF BUILDINGS AND FLOORS BY THE RUNNING OF EXPLOSIVE MOTORS	100
XIV.	HEAT EFFICIENCIES	102
XV.	EXPLOSIVE ENGINE TESTING	107
XVI.	VARIOUS TYPES OF ENGINES AND MOTORS	112
XVII.	UNITED STATES PATENTS ON GAS, GASOLINE AND OIL ENGINES, AND THEIR ADJUNCTS, SINCE 1875	269

GAS, GASOLINE, AND OIL ENGINES.

CHAPTER I.

INTRODUCTORY.

Much attention is now being given by mechanical engineers to the economical results developed in the working of gas, gasoline, and oil engines for higher powers from producer and other cheap gases. In an economical sense, for small powers steam has been left far behind.

It now becomes a question as to how to adapt the design of the new prime-movers to a wider range of usefulness.

The best steam engines now made run with a consumption of about one and three-fourth pounds of coal per horse-power per hour; while from two and one-half to seven pounds is the cost of power per horse-power per hour in the various kinds of engines now in use. This only covers the cost of fuel; the attendance required in the use of small steam power is often far greater in cost than the fuel.

When we come to require the larger powers by steam, in which economy may be obtained by compounding and condensing, the facility for obtaining the requisite water-supply is often a bar to its use. The direction in which lies the line of improvement for larger powers with the utmost economy is as yet a mooted point of discussion in explosive motor engineering.

The expansion of single-cylinder dimensions involves practical problems in the progress of ignition of the charge, as well as the thoroughness of mixture of the combustibles, and the interference of the products of the previous combustion by producing areas of imperfect or non-combustion or "stratification," as treated in foreign publications.

The enlargement of cylinder area is a source of engine-friction economy, while, on the contrary, the multiplication of cylinders involves numbers and complexity of moving parts, which go to make disparity between the indicated and brake horse-power, which is the measure of machine efficiency.

An impulse at every stroke, so desirable in an explosive motor and so satisfactorily carried out in the steam engine in connection with the compound system, seems to have as yet no counterpart in the explosive motor. Condensation is impossible, and the trials of explosion at every stroke in European engines have not proved satisfactory in service, and in order to accomplish the desired result resort has been had to duplicating single-acting cylinders. This class of explosive engines seems to fill the bill in effect; yet the complication of a two-cylinder engine as a moving mechanism must compete with a single-cylinder steam engine.

The principal types of explosive motors seem to have gone through a series of practical trials during the past thirty years, which have finally reduced the principles of action to a few permanent forms in the design of motors, that show by long-continued use the prospect of their staying qualities and their efficiency; for these will no doubt be the principal points in the final judgment of purchasers in the selection of motive power. For a gas, gasoline, or oil explosive power to approximate an ideal standard as a prime-mover, it should be simple in design, not liable to get out of order, the parts must be readily accessible, the ignition of the charge must be positive, the governing close, the engine must run quietly, and must be durable and economical in the use of fuel. These points of excellence have been striven for by many designers and builders, with varying success. But to get the entire combination without the sacrifice of some good point is not an easy matter.

But for all, the internal combustion engine has come seemingly like an avalanche of a decade; but it has come to stay, to take its well-deserved position among the powers for aiding labor.

HISTORICAL.

Although the ideal principle of explosive power was conceived some two hundred years since, and experiments made with gunpowder as the explosive element, it was not until the last years of the eighteenth century that the idea took a patentable shape, and not until about 1826 (Brown's gas-vacuum engine) that a further progress was made in England by condensing the products of combustion by a jet of water, thus creating a partial vacuum.

Brown's was probably the first explosive engine that did real work. It was clumsy and unwieldy and was soon relegated to its place among the failures of previous experiments. No approach to active explosive effect in a cylinder was reached in practice, although many ingenious designs were described, until about 1838 and the following years. Barnett's engine in England was the first attempt to compress the charge before exploding. From this time on to about 1860 many patents were issued in Europe and a few in the United States for gas engines, but the progress was slow, and its practical introduction for ordinary power purposes came with spasmodic effect and low efficiency.

From 1860 on, practical improvement seems to have been made and the Lenoir motor was produced in France and brought to the United States. It failed to meet expectations, and was soon followed by further improvements in the Hugon motor in France (1862) followed by Beau de Rocha's four-cycle idea, which has been slowly developed through a long series of experimental trials by different inventors. In the hands of Otto and Langdon a further progress was made, and numerous patents were issued in England, France, and Germany, and followed up by an increasing interest in the United States with a few patents.

From 1870 on, improvements seem to have advanced at a steady rate, and largely in the valve gear and precision of governing for variable load.

The early idea of the necessity of slow combustion was a great drawback in the advancement of efficiency, and the suggestions of de Rocha, in 1862, did not take root as a prophetic truth until many failures and years of experience had taught the fundamental axiom that rapidity of action in both combustion and expansion was the basis of success in explosive motors.

With this truth and the demand for small and safe prime-movers, the manufacture of gas engines increased in Europe and America at a more rapid rate, and improvements in perfecting the details of this cheap and efficient prime-mover have finally raised it to the dignity of a standard motor and a rival of the steam engine for small and intermediate powers, with a prospect of largely increasing its individual units to the hundred, if not to the thousand, horse-power in a single engine. The efforts of Otto, in Germany, in developing the four-cycle type, have given his name to the compression engine, which is a well-deserved tribute to genius.

The eight hundred patents issued during the past thirty years in the United States have had a simplifying tendency in construction, and have brought the efficiency of the gas, gasoline, and oil explosive engines to their present high degree of economy and widespread adoption as a prime-mover.

In this work the various changes that the gas engine has undergone in design in its European development are not considered essential to American readers, as the best European ideas have been adapted here with the spirit of American enterprise in perfecting details of construction and the application of the best material for wear in all its parts; so that in representing as many engines of American manufacture as can be obtained, the whole range of practical design will be sufficiently illustrated and described as to give a fairly good explanation of their operation to the general reader and to the users of American gas, gasoline, and oil engines.

The intense interest manifested by American engineers and inventors in the new motive power is well shown in the progress of patents issued during the past twenty years. In 1875 3 patents were issued in the United States for gas engines; 1876, 3 patents; 1877, 5 patents; 1878, 1 patent; 1879, 6 patents; 1880-81, 7 each year; 1882, 14 patents; 1883 was a booming year in gas-engine invention—no less than 40 patents were issued that year, followed by 36 patents in 1884 and 40 patents in 1885, 46 in 1886, 25 in 1887, 31 in 1888, and 58 in 1889, with an average of about 80 patents per annum during the past seven years.

The application of the gasoline motor to marine propulsion and to the horseless vehicle, the tricycle and bicycle, has had a most stimulating effect in adapting ways and means for applying this power to so many uses. Even aerial navigation has come in for its share in motor patents.

Although the denser population of Europe claims a very large representation of explosive motors in use for all purposes, the manufacture in the United States is fast forging ahead in its output of explosive motor power, for there are now no less than one hundred establishments in the United States engaged in their manufacture, and the motors in operation number many thousands. Their safety and easy management as well as their economy have made in their adoption as agricultural helpers a marvellous inroad on the old-fashioned hand and horsepower. Their later developed adaptability as a means for generating electricity for electric lighting and transmission of power is fast expanding the use of lighting and power in fields that the higher cost of small steam power had precluded. Thus the incentive to invention has been the father to a fast-growing industry, that has and will continue to ameliorate the labor of our small industries by the supply of small, reliable, and cheap power for all purposes; and present indications are that the explosive motor will become a prominent source of power for street railways, for larger sizes of vessels than heretofore used, and for stationary power, rivalling steam power of but a few years since.

CHAPTER II.

THEORY OF THE GAS AND GASOLINE ENGINE.

THE laws controlling the elements that create a power by their expansion by heat due to combustion, when properly understood, become a matter of computation in regard to their value as an agent for generating power in the various kinds of explosive engines.

The method of heating the elements of power in explosive engines greatly widens the limits of temperature as available in other types of heat engines. It disposes of many of the practical troubles of hot-air and even of steam engines, in the simplicity and directness of application of the elements of power. In the explosive engine the difficulty of conveying heat for producing expansive effect by convection is displaced by the generation of the required heat within the expansive element and at the instant of its useful work. The low conductivity of heat to and from air has been the great obstacle in the practical development of the hot-air engine; while, on the contrary, it has become the source of economy and practicability in the development of the internal-combustion engine.

The action of air, gas, and the vapors of gasoline and petroleum oil, whether singly or mixed, is affected by changes of temperature, practically in nearly the same ratio; but when the elements that produce combustion are interchanged in confined spaces, there is a marked difference of effect. The oxygen of the air, the hydrogen and carbon of a gas, or vapor of gasoline or petroleum oil are the elements that by combustion produce heat to expand the nitrogen of the air and the watery vapor produced by the union of the oxygen in the air and the hydrogen in the gas, as well as also the monoxide and carbonic-acid gas that may be formed by the union of the carbon of gas or vapor with part of the oxygen in the air.

The various mixtures as between air and gas, or air and vapor, with the proportion of the products of combustion left in the cylinder from a previous combustion, form the elements to be considered in estimating the amount of pressure that may be obtained by their combustion and expansive force.

The phenomena of the brilliant light and its accompanying heat at the moment of explosion have been witnessed in the experiments of Dugald Clerk in England, the illumination lasting throughout the stroke; but in regard to time in a four-cycle engine, the incandescent state exists only one-quarter of the running time. Thus the time interval, together with the non-conductibility of the gases, makes the phenomena of a high-temperature combustion within the comparatively cool walls of a cylinder a practical possibility.

The natural laws, long since promulgated by Boyle, Gay Lussac, and others, on the subject of the expansion and compression of gases by force and by heat, and their variable pressures and temperatures when confined, are conceded to be practically true and applicable to all gases, whether single, mixed, or combined.

The law formulated by Boyle only relates to the compression and expansion of gases without a change of temperature, and is stated in these words:

If the temperature of a gas be kept constant, its pressure or elastic force will vary inversely as the volume it occupies.

It is expressed in the formula $P \times V = C$, or pressure $\times$ volume = constant. Hence, $\dfrac{C}{P} = V$ and $\dfrac{C}{V} = P$.

Thus the curve formed by increments of pressure during the expansion or compression of a given volume of gas without change of temperature is designated as the isothermal curve, in which the volume multiplied by the pressure is a constant value in expansion, and inversely the pressure divided by the volume is a constant value in compressing a gas.

But as compression and expansion of gases require force for its accomplishment mechanically, or by the application or abstraction of heat chemically, or by convection, a second condition becomes involved, which was formulated into a law of thermodynamics by Gay Lussac under the following conditions:

A given volume of gas under a free piston expands by heat and contracts by the loss of heat, its volume causing a proportional movement of a free piston equal to $\frac{1}{273}$ part of the cylinder volume for each degree Centigrade difference in temperature, or $\frac{1}{460}$ part of its volume for each degree Fahrenheit.

With a fixed piston (constant volume), the pressure is increased or decreased by an increase or decrease of heat in the same proportion of $\frac{1}{273}$ part of its pressure for each degree Centigrade, or $\frac{1}{460}$ part of its pressure for each degree Fahrenheit change in temperature.

This is the natural sequence of the law of mechanical equivalent, which is a necessary deduction from the principle that nothing in nature can be lost or wasted, for all the heat that is imparted to or abstracted from a gaseous body must be accounted for, either as heat or its equivalent transformed into some other form of energy.

In the case of a piston moving in a cylinder by the expansive force of heat in a gaseous body, all the heat expended in expansion of the gas is turned into work; the balance must be accounted for in absorption by the cylinder or radiation.

This theory is equally applicable to the cooling of gases by abstraction of heat or by cooling due to expansion by the motion of a piston.

The denominators of these fractions represent the absolute zero of cold below the freezing-point of water, and reads $-273°$ C. or $-492.66° = -460.66°$ F. below zero; and these are starting-points of reference in computing the heat expansion in gas engines.

According to Boyle's law, called the first law of gases, there are but two characteristics of a gas and their variations to be considered, viz., volume and pressure; while by the law of Gay Lussac, called the second law of gases, a third is added, consisting of the value of the absolute temperature, counting from absolute zero to the temperatures at which the operations take place.

The ratio of the variation of the three conditions—volume, pressure, and heat from the absolute zero temperature—has a certain rate, in which the volume multiplied by the pressure and the product divided by the absolute temperature equals the ratio of expansion for each degree.

The expansion of a gas $\frac{1}{273}$ of its volume for every degree Centigrade, added to its temperature, is equal to the decimal .00366, the coefficient of expansion for Centigrade units. To any given volume of a gas, its expansion may be computed by multiplying the coefficient by the number of degrees, and by reversing the process the degree of acquired heat may be obtained approximately. These methods are not strictly in conformity with the absolute mathematical formula, because there is a small increase in the increment of expansion of a dry gas, and there is also a slight difference in the increment of expansion due to moisture in the atmosphere and to the vapor of water formed by the union of the hydrogen and oxygen in the combustion chamber of explosive engines.

The ratio of expansion on the Fahrenheit scale is derived from the absolute temperature below the freezing-point of water (32°) to correspond with the Centigrade scale; therefore $\frac{1}{492.66} = .0020297$, the ratio of expansion from 32° for each degree rise in temperature on the Fahrenheit scale.

As an example, if the temperature of any volume of air or

gas at constant volume is heated, say from 60° to 2000° F., the increase in temperature will be 1,940°. Then by the formula:

Ratio × acquired temp. × initial pressure = the gauge pressure; and .0020297 × 1940 × 14.7 = 57.88 lbs.

By another formula, a convenient ratio is obtained by $\frac{\text{absolute pressure}}{\text{absolute temp.}}$ or $\frac{14.7}{492.66} = .029838$; then, using the difference of temperature as before, .029838 × 1940 = 57.88 lbs. pressure.

By another formula, leaving out a small increment due to specific heat at high temperatures:

I. $\frac{\text{Atmospheric pressure} \times \overline{\text{absolute temp.} + \text{acquired temp.}}}{\text{Absolute temp.} + \text{initial temp.}} = $ absolute pressure due to the acquired temperature, from which the atmospheric pressure is deducted for the gauge pressure. Using the foregoing example, we have $\frac{14.7 \times \overline{460.66 + 2000}}{460.66 + 60}$ = 69.47 − 14.7 = 54.77, the gauge pressure, 460.66°, being the absolute temperature for zero Fahrenheit.

For obtaining the volume of expansion of a gas from a given increment of heat, we have the approximate formula:

II. $\frac{\text{Volume} \times \overline{\text{absolute temp.} + \text{acquired temp.}}}{\text{Absolute temp.} + \text{initial temp.}} = $ heated volume.

In applying this formula to the foregoing example, the figures become:

$$1 \times \frac{460.66 + 2000}{460.66 + 60} = 4.7341 \text{ volumes.}$$

From this last term the gauge pressure may be obtained as follows:

III. 4.7341 × 14.7 = 69.59 lbs. absolute − 14.7 lbs. atmospheric pressure = 54.89 lbs. gauge pressure; which is the theoretical pressure due to heating air in a confined space, or at constant volume from 60° to 2000° F.

By inversion of the heat formula for absolute pressure we

have the formula for the acquired heat, derived from combustion at constant volume from atmospheric pressure to gauge pressure plus atmospheric pressure as derived from Example I., by which the expression—

$$\frac{\text{absolute pressure} \times \text{absolute temp.} + \text{initial temp.}}{\text{initial absolute pressure}}$$

= absolute temperature + temperature of combustion, from which the acquired temperature is obtained by subtracting the absolute temperature.

Then, for Example 1, $\dfrac{69.47 \times 460.66 + 60}{14.7} = 2460.66$, and $2460.66 - 460.66 = 2000°$, the theoretical heat of combustion. The dropping of terminal decimals makes a small decimal difference in the result in the different formulas.

By Joule's law of the mechanical equivalent of heat, whenever heat is imparted to an elastic body, as air or gas, energy is generated and mechanical work produced by the expansion of the air or gas. When the heat is imparted by combustion within a cylinder containing a movable piston, the mechanical work becomes a measurable amount by the observed pressure and movement of the piston.

The heat generated by the explosive elements and the expansion of the non-combining elements of nitrogen and water vapor that may have been injected into the cylinder as moisture in the air, and the water vapor formed by the union of the oxygen of the air with the hydrogen of the gas, all add to the energy of the work from their expansion by the heat of internal combustion.

As against this, the absorption of heat by the walls of the cylinder, the piston, and cylinder head or clearance walls, becomes a modifying condition in the force imparted to the moving piston.

It is found that when any explosive mixture of air and gas or hydrocarbon vapor is fired, the pressure falls far short of the pressure computed from the theoretical effect of the heat produced, and from gauging the expansion of the contents of a cylinder.

It is now well known that in practice the high efficiency which is promised by theoretical calculation is never realized; but it must always be remembered that the heat of combustion is the real agent, and that the gases and vapors are but the medium for the conversion of inert elements of power into the activity of energy by their chemical union.

The theory of combustion has been the leading stimulus to large expectations with inventors and constructors of explosive motors; its entanglement with the modifying elements in practice has delayed the best development in construction, and as yet no positive design of best form or action seems to have been accomplished.

One of the most serious entanglements in the practical development of pressure due to the theoretical computations of the pressure value of the full heat is probably caused by imparting the heat of the fresh charge to the balance of the previous charge that has been cooled by expansion from the maximum pressure to near the atmospheric pressure of the exhaust. The retardation in the velocity of combustion of perfectly mixed elements is now well known from experimental trials with measured quantities; but the principal difficulty in applying these conditions to the practical work of an explosive engine where a necessity for a large clearance space cannot be obviated, is in the inability to obtain a maximum effect from the imperfect mixture and the mingling of the products of the last explosion with the new mixture, which produces a clouded condition that makes the ignition of the mass irregular or chattering, as observed in the expansion lines of indicator cards.

Stratification of the mixture has been claimed as taking place in the clearance chamber of the cylinder; but this is not satisfactory, in view of the vortical effect of the violent injection of the air and gas or vapor mixture. It certainly cannot become a perfect mixture in the time of a stroke of a high-speed motor of the two-cycle class. In a four-cycle engine, making 300 revolutions per minute, the injection and compression take place in one-fifth of a second—far too short a time for a perfect infusion of the elements of combustion.

In an experimental way, the velocity of explosion of a perfect mixture of 2 volumes of hydrogen and 1 volume of oxygen has been found to approximate 65 feet per second; and for equal volumes of hydrogen and oxygen, 32 feet per second; with 1 volume coal gas to 5 volumes air, 3¼ feet per second; 1 volume coal gas to 6 volumes of air, 1 foot per second; and with an increasing proportion of air, 10 to 9 inches per second. These velocities were obtained in tubes fired at one end only. When the ignition was made in a closed tube, so that compression was produced by the expansion from combustion, the velocity was largely increased; and with compressed mixtures, a great increase of velocity was obtained over the above-stated figures.

The different values of time, pressure, and computed heat of combustion are shown in Table 1, and graphically compared in the diagram Fig. 1.

The mixtures were Glasgow, Scotland, coal gas and air. The table and the diagram (Fig. 1) make an excellent study of the conditions of time and pressure, as well as also of the control of the work of a gas engine, by varying the proportions of the mixture.

TABLE I.—EXPLOSION AT CONSTANT VOLUME IN A CLOSED CHAMBER.

Diagram curve Fig. 1.	Mixture injected.	Time of explosion. Second.	Gauge pressure. Pounds per square inch.	Computed temperature, Fahr.
a	1 volume gas to 13 volumes air.	0.28	52	1,916°
b	1 " " " 11 " "	0.18	63	2,309
c	1 " " " 9 " "	0.13	69	2,523
d	1 " " " 7 " "	0.07	89	3,236
e	1 " " " 5 " "	0.05	96	3,484

The irregularity of the explosive curves in the diagram is fair evidence of imperfect diffusion of the gas and air mixture

THEORY OF THE GAS AND GASOLINE ENGINE. 15

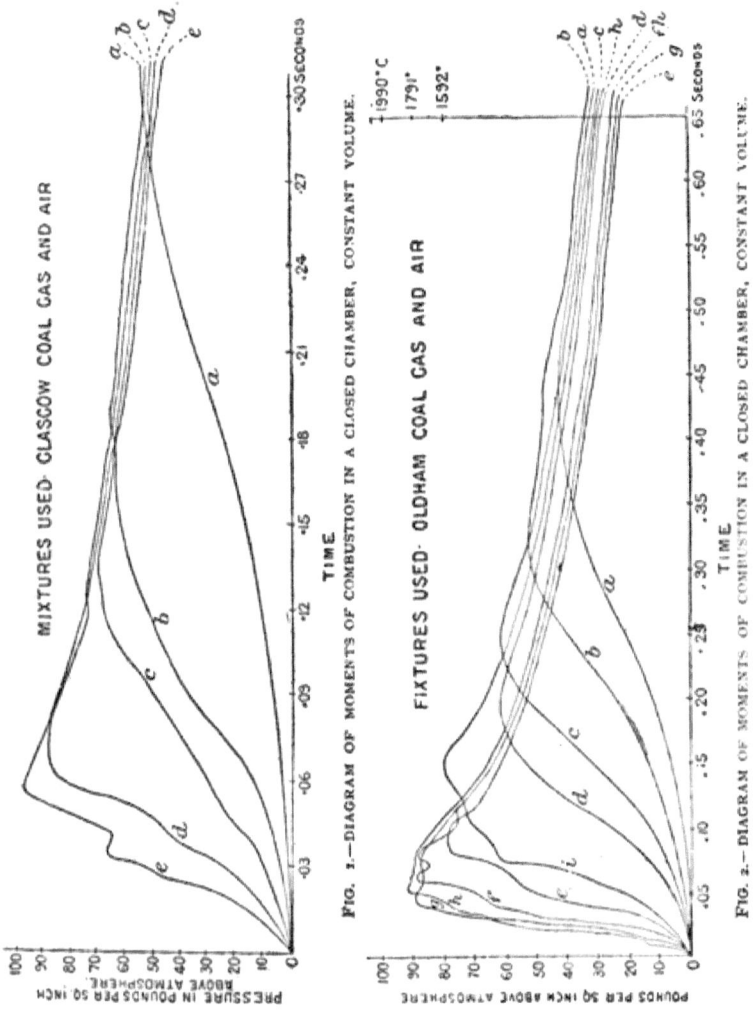

FIG. 1.—DIAGRAM OF MOMENTS OF COMBUSTION IN A CLOSED CHAMBER, CONSTANT VOLUME.

FIG. 2.—DIAGRAM OF MOMENTS OF COMBUSTION IN A CLOSED CHAMBER, CONSTANT VOLUME.

at the moment of combustion, assuming that the indicator was in perfect action.

Experiments with mixtures of coal gas and air made at Oldham, England, show a slight variation of effect, which is probably due to different proportions of hydrogen and carbon in the Oldham gas, with the same elements in the Glasgow gas. In Table 2 the injection temperature is given, which in itself is not important further than as a basis for computing the theoretical temperature of combustion.

A record of the hygrometric state of the atmosphere in its extremes would be valuable in showing the variation in explosive effect due to the vapor of water derived from the air under different hygrometric conditions.

TABLE II.—EXPLOSION AT CONSTANT VOLUME IN A CLOSED CHAMBER.

Diagram curve Fig. 7.	Mixture injected.	Temp. of injection, Fahr.	Time of explosion. Second.	Observed gauge pressure. Pounds.	Computed temp., Fahr.
a	1 volume gas to 14 volumes air.	64°	0.45	40.	1,483°
b	1 " " " 13 " "	51	0.31	51.5	1,859
c	1 " " " 12 " "	51	0.24	60.	2,195
d	1 " " " 11 " "	51	0.17	61.	2,228
e	1 " " " 9 " "	62	0.08	78.	2,835
f	1 " " " 7 " "	62	0.06	87.	3,151
g	1 " " " 6 " "	51	0.04	90.	3,257
h	1 " " " 5 " "	51	0.055	91.	3,293
i	1 " " " 4 " "	66	0.16	80.	2,371

In an examination of the times of explosion and the corresponding pressures in both tables, it will be seen that a mixture of 1 part gas to 6 parts air is the most effective and will give the highest mean pressure in a gas engine.

In this diagram the undulations of the rising curves due to irregular firing of the mixture are well marked. There is a limit to the relative proportions of illuminating gas and air mixture that is explosive, somewhat variable, depending upon the proportion of hydrogen in the gas. With ordinary coal gas, 1 of gas to 15 parts air; and on the lower end of the scale,

1 volume of gas to 2 parts of air are non-explosive. With gasoline vapor the explosive effect ceases at 1 to 16, and a saturated mixture of equal volumes of vapor and air will not explode, while the most intense explosive effect is from a mixture of 1 part vapor to 9 parts air. In the use of gasoline and air mixtures from a carburetter, the best effect is from 1 part saturated air to 8 parts free air.

CHAPTER III.

UTILIZATION OF HEAT AND EFFICIENCY IN GAS ENGINES.

The utilization of heat in any heat engine has long been a theme of inquiry and experiment with scientists and engineers, for the purpose of obtaining the best practical conditions and construction of heat engines that would represent the highest efficiency or the nearest approach to the theoretical value of heat, as measured by empirical laws that have been derived from experimental researches relating to its ultimate value. It is well known that the steam engine returns only from 12 to 18 per cent. of the power due to the heat generated by the fuel, about 25 per cent. of the total heat being lost in the chimney, the only use of which is to create a draught for the fire; the balance, some 60 per cent., is lost in the exhaust and by radiation. The problem of utmost utilization of force in steam has nearly reached its limit.

The internal-combustion system of creating power is comparatively new in practice, and is but just settling into definite shape by repeated trials and modification of details, so to give somewhat reliable data as to what may be expected from the rival of the steam engine as a prime-mover.

For small powers, the gas, gasoline, and petroleum oil engine is forging ahead at a rapid rate, filling the thousand wants of manufacture and business for a power that does not require expensive care, that is perfectly safe at all times, that can be used in any place in the wide world to which its concentrated fuel can be conveyed, and that has eliminated the constant handling of crude fuel and water.

The utilization of heat in a gas engine is mainly due to the manner in which the products entering into combustion are distributed in relation to the movement of the piston.

In the two-cycle engine, the gas or vapor and air mixtures are drawn in during a part of the stroke, fired, expanded with the motion of the piston, and exhausted by the return stroke. The proportions of the indraught to the stroke of the piston, and the volume of the clearance or combustion chamber, as it is usually called, have been subject to a vast amount of experi-

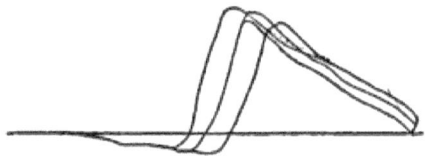

FIG. 3.—LENOIR TYPE.

ment and practical trial, in an endeavor to bring the heat value of their power up to its highest possible limit.

To this class belonged some of the earlier gas engines; their indicator cards have a typical representation in Fig. 3.

The earlier engines of this class used as high as 96 cubic feet of illuminating gas per horse-power per hour. The consumption of gas fell off by improvements to 70 cubic feet, and finally has dropped to 44 and to 36 cubic feet per indicated horse-power per hour.

The efficiency of this class of gas engines has seldom reached 20 per cent. of the heat value of the gas used, while in the compression or four-cycle engines there are possibilities of 35 per cent. The total efficiency of the gas or vapor entering into combustion in an internal-heat engine is variable, depending upon its constituent-combining elements and the degree of temperature produced. The efficiency due to heat only varies as the difference between the initial temperature of the explosive mixture and the temperature of combustion; and as this varies in actual practice from $1400°$ to $2500°$ F., then the reciprocal of the absolute heat of the initial charge, divided by the assumed heat of combustion, would represent the total efficiency. The formula $\frac{H - H^1}{H}$ represents this condition, so that if the operation of the heat cycle was between 60° and 1,400° F., the equation would be: $\frac{60 + 460}{1400 + 460} = .279$ and $1 - .279 = .72$ per cent. But this cannot represent a working cycle from the change in the specific heat of the gaseous contents of a cylinder while undergoing expansion by the movement of a piston.

The specific heat of air at constant volume is .1685, and at constant pressure is .2375. Their ratio $\frac{.2375}{.1685} = 1.408$. The ratios of the other elements entering into combustion in a gas engine are slightly less than for air; but the ratio for air is near enough for all practical operations. The formula for the application of the condition of work with complete expansion is: $1 - 1.408 \frac{H^1}{H}$; or, as for above example, $1 - 1.408 \frac{60 + 460}{1400 + 460} = .3928$, and $1 - .3928 = .6071$, or 60 per cent.

As the temperature cannot be utilized for work from the excess of heat in the products of combustion when the expansion has reached the atmospheric line, then the practical amount of expansion and the heat of combustion at the point of exhaust must be considered. In practice, the measured heat of the exhaust at atmospheric pressure, plus the additional heat due to the terminal pressure, becomes a factor in the equation; and, assuming this to be 950° F. in a well-regulated motor, the equation for the above example becomes: $1 - 1.408 \times \frac{950 - 460}{1400 - 460} = \frac{490}{940} = .521 \times 1.408 = .733$, and $1 - .733 = .26$, or an efficiency of 26 per cent. The greater difference in temperature, other things being equal, the greater the efficiency.

In this way efficiencies are worked out through intricate formulas for a variety of theoretical and unknown conditions of combustion in the cylinder: ratios of clearance and cylinder volume, and the uncertain condition of the products of combustion left from the last impulse and the wall temperature. But they are of but little value, except as a mathematical inquiry as to possibilities. The real commercial efficiency of a gas or gasoline engine depends upon the volume of gas or liquid at some assigned cost, required per actual brake horsepower per hour, in which an indicator card should show that the mechanical action of the valve gear and ignition was as perfect as practicable, and that the ratio of clearance, space,

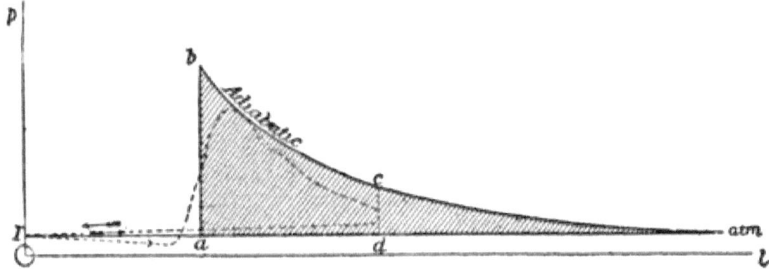

FIG. 4.—COMPARATIVE CARD.

and cylinder volume gave a satisfactory terminal pressure and compression—the difference between the power figured from the indicator card and the brake power being the friction loss of the engine.

In practice, the heat value of the gas per cubic foot may vary from 30 per cent. with illuminating and natural gases to 75 or 80 per cent. as between good illuminating gas and Dowson gas; then, in order that a given size engine should maintain its rating, a larger volume of a poorer gas should be swept through the cylinder. This requires adjustment of the areas in all the valves to give an explosive motor its highest efficiency for the kind of fuel that is to be used.

The practical effect of the work done by the half-cycle in the earlier type of the two-cycle engine is graphically shown in Fig. 4, in which i, d represents the stroke of the piston; the dotted line, the indicator card; and the space in the lines, a, b, c, d, the ideal diagram of a perfect gas exhausting at the point d, in its incomplete adiabatic expansion. In the valuation of such a card, the depression of the indraught below the atmospheric line and the pressure of the exhaust line should have due consideration as negative quantities to be deducted from the pressure values above the atmospheric line. This class of engines is fast becoming obsolete as a type.

In four-cycle engines the efficiencies are greatly advanced by compression, producing a more complete infusion of the mixture of gas or vapor and air, quicker firing, and far greater pressure than is possible with the two-cycle type just described.

In the practical operation of the gas engine during the past fifteen years, the gas-consumption efficiencies per indicated horse-power have gradually risen from 17 per cent. to a maximum of 28 per cent. of the theoretical heat, and this has been done chiefly through a decreased combustion chamber and increased compression—the compression having gradually increased in practice from 30 lbs. per square inch to above 80; but there seems to be a limit to compression, as the efficiency ratio decreases with the increase in compression.

It has been shown that an ideal efficiency of 33 per cent. for 38 lbs. compression will increase to 40 per cent. for 66 lbs., and 43 per cent. for 88 lbs. compression. On the other hand, greater compression means greater explosive pressure and greater strain on the engine structure, which will probably retain in future practice the compression between the limits of 40 and 60 lbs.

In experiments made by Dugald Clerk with a combustion chamber equal to 0.6 of the space swept by the piston, with a compression of 38 lbs., the consumption of gas was 24 cubic feet per indicated horse-power per hour. With 0.4 compression space and 61 lbs. compression, the consumption of gas was 20 cubic feet per indicated horse-power per hour; and with

0.34 compression space and 87 lbs. compression, the consumption of gas fell to 14.8 cubic feet per indicated horsepower per hour—the actual efficiencies being respectively 17, 21, and 25 per cent. This was with a Crossley four-cycle engine.

In Fig. 5 is represented an ideal card of the work of a perfect compression cycle in which the gases are compressed. Additional pressure is instantly developed by combustion or heat at constant volume, and then allowed to expand to atmospheric

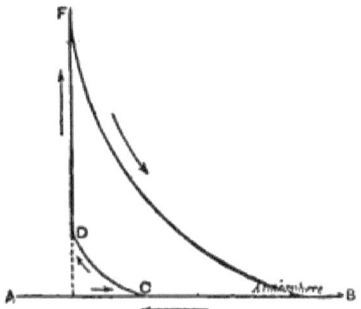

FIG. 5.—DIAGRAM OF A PERFECT CYCLE WITH COMPRESSION.

pressure—the curves of compression and expansion being adiabatic, as for a dry gas.

In this diagram the lines follow Carnot's cycle, in which the whole heat energy is represented in work. The piston stroke commencing at O, compression completed at D, pressure augmented from D to F, expansion doing work from F to B, and exhausting along the atmospheric line B A. The gases in this case expand till their pressure falls to the atmospheric line, and their whole energy is supposed to be utilized. In this imaginary cycle, no heat is supposed to be lost by absorption of walls of a cylinder or by radiation, and no back pressure during exhaust, or friction, are taken into account.

The efficiencies in regard to power in a heat engine may be divided into four kinds, of which·

I. The first is known as the *maximum theoretical efficiency* of a perfect engine (represented by the lines in the indicator diagram, Fig. 5). It is expressed by the formula $\frac{T_1 - T_0}{T_1}$ and shows the work of a perfect cycle in an engine working between the received temperature + absolute temperature (T_1) and the initial atmospheric temperature + absolute temperature (T_0).

II. The second is the *actual heat efficiency*, or the ratio of the heat turned into work to the total heat received by the engine. It expresses the *indicated horse-power*.

III. The third is the ratio between the second or *actual heat efficiency* and the first or *maximum theoretical efficiency* of a perfect cycle. It represents the greatest possible utilization of the power of heat in an internal-combustion engine.

IV. The fourth is the *mechanical efficiency*. This is the ratio between the actual horse-power delivered by the engine through a dynamometer or measured by a brake (brake horse-power), and the indicated horse-power. The difference between the two is the power lost by engine friction.

CHAPTER IV.

RETARDED COMBUSTION AND WALL-COOLING.

SOME of the serious difficulties in practically realizing the condition of a perfect cycle in an internal-combustion engine are shown in the diagram Fig. 6, taken from an English Otto

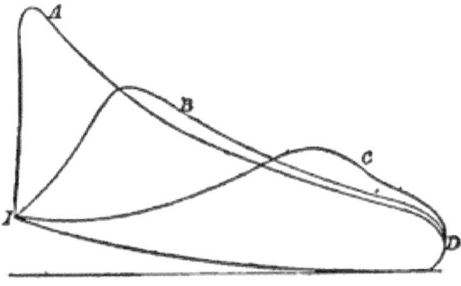

FIG. 6.—VARIABLE CARD.

gas engine, in which the cooling effect of the walls are shown by the lagging of the explosion curve, by the missing of several explosions when the cylinder walls have been unduly cooled by the water-jacket. The same delay is experienced in starting a gas engine. The indicator card I A D representing the normal condition of constant work in the cylinder; the curve I B D an interruption of explosions for several revolutions; and I C D a still longer interruption in the explosions with the engine in continuous motion.

In an experimental investigation of the efficiency of a gas engine under variable piston speeds made in France, it was found that the useful effect increases with the velocity of the piston—that is, with the rate of expansion of the burning gases with mixtures of uniform volumes; so that with the variations of time of complete combustion at constant pressure, as illustrated on page 15, and the variations due to speed, in a way compensate in their efficiencies. The dilute mixture, being slow burning, will have its time and pressure quickened by increasing the speed.

TABLE V.—TRIAL EFFICIENCIES DUE TO INCREASED PISTON SPEED.

$$\text{Efficiency} = \frac{\text{work of indicator diagram}}{\text{theoretical work.}}$$

Mixtures.	Time of explosion. Second	Piston speed. Foot per second.	Computed work diagram. Foot-pounds.	Theoretical work of the gas. Foot-pounds.	Efficiency.
1 volume coal gas to 9.4 volumes air (.1093 cubic feet mixture)...............	.53	1.181	70.8	4917	1.44
1 volume coal gas to 9.4 volumes air....	.40	1.64	85.3	4917	1.70
1 " " " 9.4 " "	.25	3.01	105.5	4917	2.10
1 " " " 9.4 " "	.16	4.55	125.8	4917	2.60
1 " " " 6.33 " " (.073 cubic feet mixture)...............	.15	5.57	127.2	4793	2.60
1 volume coal gas to 6.33 volume air....	.09	9.51	289.9	4793	6.00
1 " " " 6.33 " "	.06	14.1	364.4	4793	7.50

These trials give unmistakable evidence that the useful effect increases with the velocity of the piston—that is, with the rate of expansion of the burning gases.

The time necessary for the explosion to become complete and to attain its maximum pressure depends not only on the composition of the mixture, but also upon the rate of expansion.

This has been verified in experiments with the Kane-Pennington motor, at speeds from 500 to 1,000 revolutions per minute, or piston speeds of from 16 to 32 feet per second.

The increased speed of combustion due to increased piston speed is a matter of great importance to builders of gas engines, as well as to the users, as indicating the mechanical direction of improvements to lessen the wearing strain due to high speed and to lighten the vibrating parts with increased strength, in order that the balancing of high-speed engines may be accomplished with the least weight.

From many experiments made in Europe, it has been conclusively proved that excessive cylinder cooling by the water-jacket is a loss of efficiency.

In a series of experiments with a simplex engine in France, it was found that a saving of 7 per cent. in gas consumption per brake horse-power was made by raising the temperature of

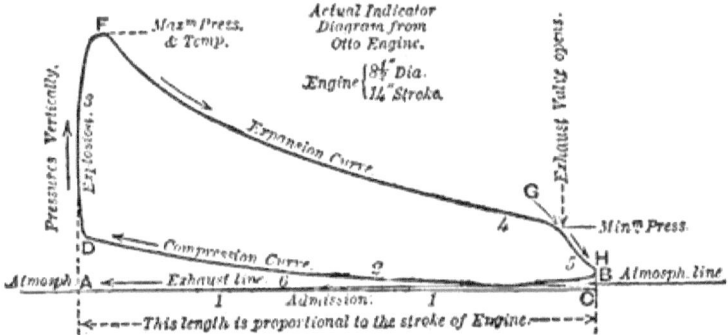

FIG. 7.—OTTO FOUR-CYCLE CARD.

the jacket water from 141° to 165° F. A still greater saving was made in a trial with an Otto engine by raising the temperature of the jacket water from 61° to 140° F.—it being 9.5 per cent. less gas per brake horse-power.

In view of the experiments in this direction, it clearly shows that in practical work, to obtain the greatest economy per effective brake horse-power, it is necessary:

1st. To transform the heat into work with the greatest rapidity mechanically allowable. This means high piston speed.

2d. To have high initial compression.

3d. To reduce the duration of contact between the hot gases and the cylinder walls to the smallest amount possible; which means short stroke and quick speed.

4th. To adjust the temperature of the jacket water to obtain the most economical output of actual power. This means water tanks or water coils, with air-cooling surfaces suitable

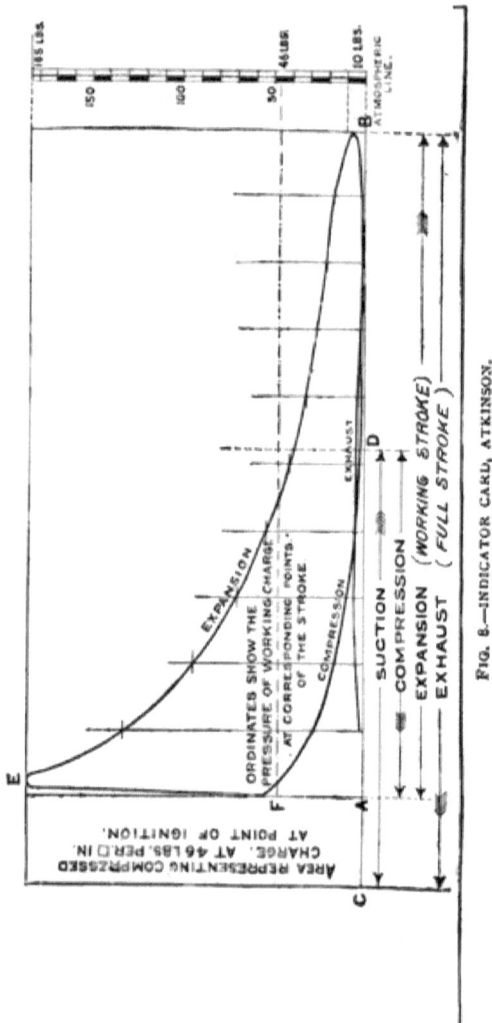

Fig. 8.—Indicator Card, Atkinson.

and adjustable to the most economical requirement of the engine.

5th. To reduce the wall surface of the clearance space or

combustion chamber to the smallest possible area, in proportion to its required volume. This lessens the loss of the heat of combustion by exposure to a large surface, and allows of a higher mean wall temperature to facilitate the heat of compression.

It will be noticed that the volumes of similar cylinders increase as the cube of their diameters, while the surface of their

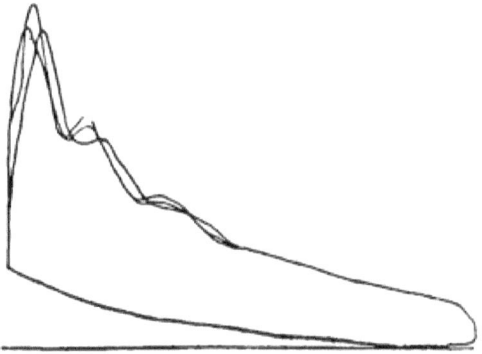

FIG. 9.—INDICATOR CARD, FULL LOAD.

cold walls varies as the square of their diameters; so that for large cylinders the ratio of surface to volume is less than for small ones. This points to greater economy in the larger engines.

The study of many experiments goes to prove that combustion takes place gradually in the gas-engine cylinder, and that the rate of increase of pressure or rapidity of firing is controlled by dilution and compression of the mixture, as well as by the rate of expansion or piston speed.

The rate of combustion also depends on the size and shape of the exploding chamber, and is increased by mechanical agitation of the mixture during combustion, and still more by the mode of firing. A small intermittent spark gives the most uncertain ignition, whereas a continuous electric spark passed through an explosive mixture, or a large flame as the shooting of a mass of lighted gas into a weak mixture, will produce rapid ignition.

The shrinkage of the charge of mixed gas and air by the union of its hydrogen and oxygen constituents by the production of the vapor of water in a gas-engine cylinder, using 1 part illuminating gas to 6.05 parts air, is a notable amount,

FIG. 10.—INDICATOR CARD, HALF LOAD.

and of the total volume of 7.05 in cubic feet. the product will be:

```
1.3714 cubic feet water vapor.
 .5714   "     "  carbonic acid.
 .0050   "     "  nitrogen derived from the gas.
4.8000   "     "    "      "      "    "   air.
─────            "     "  products of combustion.
6.7428
```

Then 7.05 cubic feet of the mixture charge will have shrunk by combustion to 6.7428 cubic feet at initial temperature, or 4.4 per cent.

This difference in the computed shrinkage at initial temperature is manifested in the reduced pressure of combustion due to the computed shrinkage, and amounts to about 2 per cent. in the mean pressure, as shown on an indicator card.

With the less rich gas, as water and Dowson gas, the shrinkage by conversion into water vapor is equal to 5.5 per cent.

In Fig. 7 is shown an actual indicator diagram from an English Otto engine, in which the sequence of operations are delineated through two of its four cycles. The curve of explosion shows that firing commenced slightly before the end of the stroke, and that combustion lagged until a moment after reversal of the stroke. The expansion line is somewhat higher than the adiabatic curve, indicating a partial combustion taking place during the stroke of the piston, and particularly

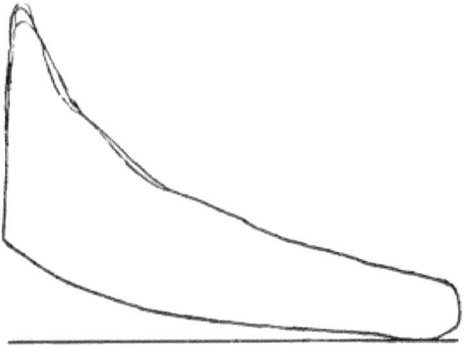

FIG. 11.—TYPICAL COMPRESSION CARD. MEAN PRESSURE, 76 LBS. PER SQUARE INCH.

manifested by the rounding-off of the apex of the card.

In Fig. 8 is represented a card from the Atkinson gas engine. The peculiar design of this engine enables the largest degree of expansion known in gas-engine practice.

Fig. 9 is a card from a compression engine, showing an irregularity in firing the charge, and probably an irregular progress of combustion by defective mixture. This card was made when running at full load, and computed at 69 lbs. mean pressure.

Fig. 10 represents a card from the same engine at half-load and lessened combustion charge. It shows the same characteristics as to irregularity, and also a lag in firing and a fitful after-combustion; but from weak mixture and interrupted firing the cooling influence of the cylinder walls has prolonged the combustion with ignition pressure. Mean pressure, about 68 lbs. per square inch.

Fig. 11 represents a typical card of our best compression

engines, with time igniter, at full load and uninterrupted firing.

Examples of indicator cards from engines in which firing commenced just before the end of the compression stroke make a rounded corner at the end of the compression curve, which is claimed to make the running of the engine smoother or without jar from the sudden increase in pressure.

CHAPTER V.

CAUSES OF LOSS AND INEFFICIENCY IN EXPLOSIVE MOTORS.

The difference realized in the practical operation of an internal-heat engine from the computed effect derived from the values of the explosive elements is probably the most serious difficulty that engineers have encountered in their endeavors to arrive at a rational conclusion as to where the losses were located and the ways and means of design that would eliminate the causes of loss and raise the efficiency step by step to a reasonable percentage of the total efficiency of a perfect cycle.

The loss of heat to the walls of the cylinder, piston, and clearance space, as regards the proportion of wall surface to the volume, has gradually brought this point to its smallest ratio in the concave piston head and globular cylinder head, with the smallest possible space in the inlet and exhaust passage. The wall surface of a cylindrical clearance space or combustion chamber of one-half its unit diameter in length is equal to 3.1416 square units, its volume but 0.3927 of a cubic unit; while the same wall surface in a spherical form has a volume of 0.5236 of a cubic unit. It will be readily seen that the volume is increased $33\frac{1}{3}$ per cent. in a spherical over a cylindrical form for equal wall surfaces at the moment of explosion, when it is desirable that the greatest amount of heat is generated and carrying with it the greatest possible pressure from which the expansion takes place by the movement of the piston.

The spherical form cannot continue during the stroke for mechanical reasons; therefore some proportion of piston stroke or cylinder volume must be found to correspond with a spherical form of the combustion chamber to produce the least loss of heat through the walls during the combustion and expansion part of the stroke.

This idea we illustrate in Figs. 12 and 13, showing how the relative volumes of cylinder stroke and combustion chamber

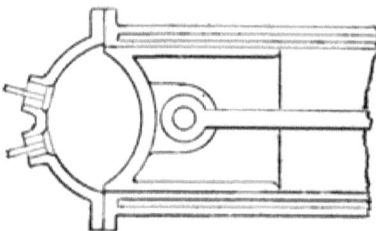

FIG. 12.—SPHERICAL COMBUSTION CHAMBER.

may be varied to suit the requirements due to the quality of the elements of combustion. In Fig. 12 the ratio may also be decreased by extending the stroke. The mean temperature of the wall surface of the combustion chamber and cylinder, as indicated by the temperatures of the circulating water, has been found to be an important item in the economy of the gas

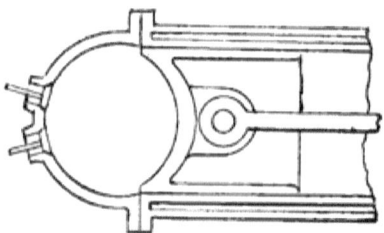

FIG. 13.—ENLARGED COMBUSTION CHAMBER.

engine. Dugald Clerk, in England, a high authority in practical work with the gas engine, found that 10 per cent. of the gas for a stated amount of power was saved by using water at a temperature in which the ejected water from the cylinder jacket was near the boiling point, and ventures the opinion that a still higher temperature for the circulating water may be used as a source of economy.

This could be made practical by elevating the water tank and adjusting the air-cooling surface, so as to maintain the inlet water at just below the boiling-point, and by the rapid circulation induced by the height of the tank above the engine and the pressure, to return the water from the cylinder jacket a few degrees above the boiling-point.

For a given amount of heat taken from the cylinder by the largest volume of circulating water, the difference in temperature between inlet and outlet of the water jacket should be the least possible, and this condition of the water circulation gives a more even temperature to all parts of the cylinder; while, on the contrary, a cold water supply, say at $60°$ F., so slow as to allow the ejected water to flow off at a temperature near the boiling-point, must make a great difference in temperature between the bottom and top of the cylinder, with a loss in economy in gas and other fuels, as well as in water, if it is obtained by measurement.

In regard to the actual consumption of water per horsepower and the amount of heat carried off by it, the study of English trials of an Atkinson, Crossley, and Griffin engine showed 62 lbs. water per indicated horse-power per hour, with a rise in temperature of $50°$ F., or 3,100 heat units were carried off in the water out of 12,027 theoretical heat units that were fed to the motor through the 19 cubic feet of gas at 633 heat units per cubic foot per hour.

Theoretically, 2,564 heat units per hour is equal to 1 horsepower. Then 0.257 of the total was given to the jacket water, 0.213 to the indicated power, and the balance, 53 per cent., went to the exhaust, radiation, and the reheating of the previous charge in the clearance and in expanding the nitrogen of the air. Other and mysterious losses, due to the unknown condition of the gases entering into and passing through the heat cycle, have been claimed and mathematically discussed by authors, which have failed to satisfy the practical side of the question, which is the main object of this work.

In a trial with the Crossley engine, 42 lbs. of water per horse-power per hour were passed through the cylinder jacket, with a rise in temperature of 128° F.—equal to 5,376 heat units to the water from 12,833 heat units fed to the engine through 20.5 cubic feet of gas at 626 heat units per cubic foot.

In this trial, 41 per cent. of the total heat was carried away in the water; 2,564 heat units being equal to one indicated horse-power per hour, then $5,376 + 2,564 = 7,940$ were directly accounted for, leaving 38 per cent. to the exhaust and other losses. As these engines were both of the compression type, and the Crossley engine having double the clearance space of the Atkinson engine, and with so great a difference in the volumes of the previous explosion held over, a just comparison of the effect of different cylinder temperatures cannot be made. The efficiencies were found, including gas used for ignition, to be for the Atkinson, 22.8 per cent.; for the Crossley, 21.2 per cent.; and for the Griffin, a double-acting engine, 19.2 per cent. of the total gas power used. The efficiency of other engines of the four-cycle compression type in Europe varies from 17 to 22 per cent., some of the lower efficiencies being claimed as due to the composition of the low-power Dowson and water gases.

An experimental test of the performance of a gas engine below its maximum load has shown a large increase in the consumption of gas per actual horse-power, with a decrease of load, as the following figures from observed trials show: An actual 12 H.P. engine at full load used 15 cubic feet of gas per horse-power per hour; at 10 H.P., $15\frac{1}{2}$ cubic feet; at 8 H.P., $16\frac{1}{2}$ cubic feet; at 6 H.P., 18 cubic feet; at 4 H.P., 21 cubic feet; at 2 H.P., 30 cubic feet of gas per actual horse-power per hour. This indicates an economy in gauging the size of a gas engine to the actual power required, in consideration of the fact that the engine friction and gas consumption for ignition are constants for all or any power actually given out by the engine.

CHAPTER VI.

ECONOMY OF THE GAS ENGINE FOR ELECTRIC-LIGHTING.

In the lighting of large dwellings or other buildings, where there is no power used for other purposes, the use of gas or gasoline engines for operating an electric generator is not only cheaper in running expenses than the steam engine, but the comparison holds good for the lighting of towns and villages at the usual cost of gas to consumers; but when the generation of producer gas can be made for such use on the premises of the electric plant and by the same persons that operate the electric plant, the saving in cost of electric-lighting is several-fold less than by direct gas-burning.

In many towns where oil producer gas is used, the cost of material used in making the gas is less than thirty-five cents per thousand feet of gas produced. In such places the labor of producing the gas for a town of say fifteen hundred inhabitants is from two to three hours per day, and in some towns, as observed by the author, three hours every other day—giving ample time for the same operator to run the electric plant in the evening, or both may be run simultaneously.

When the mere fact of the cost of gas for direct lighting and its cost for producing the same light by its use in a gas engine to run an electric generator is considered, the difference in favor of electric-lighting in preference to direct gas-lighting is most apparent.

It has been known for some years that for equal light power but about one-half the volume of gas consumed in direct lighting will produce the same amount of candle-power when used in a gas engine for generating electricity for lighting.

Again, when we leave the realm of a fixed gas and the cost of its producing-plant, the gasoline and oil engine again comes to the rescue of the fuel element for lighting, from an average cost of $7\frac{1}{2}$ cents per hour for 192 candle-power in lights by direct illumination, and $2\frac{1}{4}$ cents for the same amount of light by the use of illuminating gas consumed in a gas engine with electric generator, to one cent or less by the gasoline and oil engine for equal light.

In English trials with a Crossley engine of 54 I.H.P. running a $25\frac{1}{2}$ kilowatt generator (34 electrical H.P.), lighting 400 incandescent lamps (16 candle-power) consumed 1,130 cubic feet illuminating gas per hour, or 2.82 cubic feet of gas per lamp per hour.

The gas used was 16 candle-power at 5 cubic feet per hour. Then, if it had been used for direct lighting, it would have produced $11\frac{30}{5} = 226 - 16$-candle-power gas-lights, a little over one-half the amount of the electric light—or the efficiency of the direct light would have been but 56.5 per cent.

To show the difference between running a gas engine at full or less than full power, the same engine and generator when running with 300 incandescent lamps, 16 candle-power, used 840 cubic feet of gas per hour, and $8\frac{40}{5} = 168 - 16$ candle-power gas-lights, or 56 per cent. efficiency for direct lighting.

When the lamps were cut out to one-half or 200, the consumption of gas was 740 cubic feet per hour, equal to $7\frac{40}{5} = 148$ gas lights, with a direct gaslight efficiency of 74 per cent.—the difference in efficiency being chiefly due to the constant value of the engine and generator friction in its relation to the variable power.

Another trial with a Tangye engine of a maximum 39 I.H.P. running an 18.36 kilowatt generator (24.61 electrical H.P.), lighting 300 16-candle-power incandescent lamps, consumed 770 cubic feet illuminating gas per hour. With direct lighting, $7\frac{70}{5} = 154$ gas-lights (16 candle-power), or an efficiency of 51 per cent. for direct lighting. With 220 incandescent lamps in,

640 cubic feet of gas were consumed per hour, equal to $\frac{640}{5} =$ 128 gas-lights and a direct gaslight efficiency of $\frac{128}{220} = 58$ per cent. Again reducing to 100 lamps, 320 cubic feet of gas was used, equal to 64 gas-lights with an efficiency of 64 per cent. for direct gaslighting.

It will readily be seen by inspection of these figures that the greatest economy in gas-engine power will be found in gauging the size of a gas engine by the work it is to do when the work is a constant quantity.

In a trial by the writer of a Nash gas engine of 5 B.H.P., driving by belt a Riker 3 kilowatt bipolar generator of 120 volts, 25 ampere capacity, the engine speed was 300 revolutions and the generator 1,400 revolutions per minute; consumption of New York gas, 105 cubic feet per hour. With 50 120-volt A.B.C. lamps in circuit giving a brilliant white light of fully 16 candle-power, the actual voltage by meter was 120, amperage by meter 24, voltage and amperage perfectly steady with continuous running. By turning in resistance and reducing the voltage to 110 and the amperage to 21, the lights were still brilliant in the 50 lamps. With the lamps cut out to 40, the voltmeter vibrated 2 volts and immediately came back to 110 volts, with the amperemeter at 17. With a further and sudden cutting out the light to 20 lamps, the voltage fell to 105 with but slight vibration; amperage, 11. With 15 lamps on, the voltage crept up to 110, amperage $6\frac{1}{2}$, and with 10 lamps only the voltage vibrated for a few seconds and rested at 110, amperage $4\frac{1}{2}$. The engine seemed to answer the change of load remarkably quick, so that there was no perceptible change in speed.

The investment of local lighting-plants by the use of gas, gasoline, and oil engines in factories and large buildings in Europe has been found a great source of economy as against the direct use of municipal electric current or the direct use of gas.

The gasoline or oil engine makes a most favorable return in economy when used for local lighting as against the prevailing price charged by the operators of large steam-power installations for town and city lighting.

In a trial of eleven days by a 10 H.P. four-cycle gas engine of the Raymond vertical pattern, belted direct to a 150-light direct-current generator making 1,600 revolutions per minute, with the current measured by a recording wattmeter, giving a steady current to 90 16-candle-power lamps on a factory circuit, the total cost of gas at $1.50 per 1,000 cubic feet with lubricating oils was $20.16. The kilowatts produced by measure was 239.1 or a cost of .0844 cents per kilowatt. The price of the current by the same measure from the electric company was 20 cents per kilowatt—a saving of 57 per cent. In places where gas is $1 per 1,000 feet, the cost would have been only $5\tfrac{3}{4}$ cents per kilowatt.

In the lighting of churches the gas or gasoline engine has been found to be not only economical, but has largely contributed to the cheerful surroundings of a lighted church at less than one-half the cost of gas for direct lighting, and with no more attention in starting the engine, cleaning, etc., than required for lighting and regulating the ordinary gas lights.

CHAPTER VII.

THE MATERIAL OF POWER IN EXPLOSIVE ENGINES.

The composition of gases, gasoline, petroleum oil, and air as elements of combustion and force in explosive engines is of great importance in comparisons of heat and motor efficiencies. By reported experiments with 20-candle coal gas in the United States, by the evaporation of water at 212° F., a cubic foot was credited with 1,236 heat units; while reliable authorities range the value of our best illuminating gases at from 675 to 700 heat units per cubic foot. The specific heat of illuminating gas is much higher than for air, being for coal gas at constant pressure 0.6844 and at constant volume 0.5196, with a ratio of 1.315; while the specific heat for air at constant pressure is 0.2377, and at constant volume is 0.1688, and their ratio 1.408.

The mixtures of gas and air accordingly vary in their specific heat with ratios relative to the volumes in the mixture. The products of combustion also have a higher specific heat than air, ranging from 0.250 at constant pressure and 0.182 at constant volume, to 0.260 and 0.190 with ratios of 1.37 and 1.36.

A cubic foot of ordinary coal gas burned in air produces about one ounce of water vapor and 0.57 of a cubic foot of carbonic acid gas (CO_2). Its calorific value will average about 673 heat units per cubic foot.

A cubic foot of ordinary coal gas requires 1.21 cubic feet of oxygen, more or less, due to variation in the constituents of different grades of illuminating gases in various localities, for complete combustion.

Allowing for an available supply of 20 per cent. of oxygen

in air for complete combustion, then $1.21 \times 5 = 6.05$ cubic feet of air which is required per cubic foot of gas in a gas engine for its best work; but in actual practice the presence in the engine cylinder of the products of a previous combustion, and the fact that a sudden mixture of gas and air may not make a homogeneous combination for perfect combustion, require a larger proportion of air to completely oxidize the gas charge.

It will be seen by inspection of Table 2 that the above proportion, without the presence of contaminating elements, produces the quickest firing and approximately the highest pressure at constant volume, and that any greater or less proportion of air will reduce the pressure and the apparent efficiency of an explosive motor. There are other considerations effecting the governing of explosive engines, in which the gas element only is controlled by the governor, requiring an excess of air at the normal speed, so that an economical adjustment of gas consumption may be obtained at both above and below the normal speed.

TABLE III.—THE MATERIALS OF POWER IN EXPLOSIVE ENGINES—GASES, GASOLINE, AND PETROLEUM OILS.

Various gases, vapors, and other combustibles.	Heat units, per pound.	Heat units, per cubic foot.	Foot-pounds, per cubic foot.
Hydrogen	61,560	293.5	226,580
Carbon	14,540		
Crude petroleum, West Virginia, spec. grav. .873	18,324		
Light petroleum, Pennsylvania, spec. grav. .841	18,401		
Benzine, C_6H_6	18,448		
Gasoline	11,000		
28 candle-power illuminating gas	....	950	773,400
19 " " "	....	800	617,600
15 " " "	....	620	478,640
Water gas, American	....	185	142,820
Producer gas, English ....66 to..	....	150	115,800
Water producer gas	....	104	80,288
Ethylene olefiant gas, C_2H_4	21,430	1677	
Gasoline vapor	11,000	690	492,680
Acetylene C_2H_2	21,492	868	670,090
Natural gas, Leechburg, Pa	....	584	450,848
" " Pittsburg, Pa	....	495	382,140
Marsh gas (Methane), CH_4	23,594	1051	

The various other than coal gas used in explosive engines are NATURAL GAS, ACETYLENE, liberated by the action of water on calcium carbide; PRODUCER GAS, made by the limited action of air alone upon incandescent fuel; WATER GAS, made by the action of steam alone upon incandescent fuel; SEMI-WATER GAS, made by the action of both air and steam upon incandescent fuel—also named DOWSON GAS in England.

Natural Gas.

The constituents of natural gas varies to a considerable extent in different localities. The following is the analysis of some of the Pennsylvania wells:

NATURAL GAS CONSTITUENTS, BY VOLUME.

Constituents.	Olean, N. Y.	Pittsburg, Pa.	Leechburg, Pa.	Harvey well, Butler county.	Burns well, Butler county.
Hydrogen, H	....	22.00	4.79	13.50	6.10
Marsh gas, CH$_4$	96.50	67.00	89.65	80.11	75.44
Ethane, C$_2$H$_4$	....	5.00	4.39	5.72	18.12
Heavy hydrocarbons	1.00	1.00	.56		
Carbonic oxide, CO	.50	.60	.26	trace.	trace.
Carbonic acid, CO$_2$	...	.60	.35	.66	.34
Nitrogen, N	....	3.00			
Oxygen, O	2.00	.80			
	100.00	100.00	100.00	100.00	100.00
Heat units, cubic feet, Fah. =	....	892	1051	959	1151

Density, 0.5 to 0.55 (air 1).

The calorific value of natural gas in much of the Western gas fields is below these figures.

In experiments recorded by Brannt, "Petroleum and Its Products," with the *oil gas* as made for town lighting in many parts of the United States, of specific gravity about 0.68 (air 1), mixtures of oil gas with air had the following explosive properties:

Oil gas, volumes.	Air, volumes.	Explosive effect.
1	4.9	None.
1	5.6 to 5.8	Slight.

Oil gas, volumes.	Air, volumes.	Explosive effect.
1	6 to 6.5	Heavy.
1	7 to 9	Very heavy.
1	10 to 13	Heavy.
1	14 to 16	Slight.
1	17 to 17.7	Very slight.
1	18 to 22	None.

It will be seen that mixtures varying from 1 of gas to 6 of air, and all the way to 1 of gas to 13 of air, are available for use in gas engines for the varying conditions of speed and power regulation; and that 1 of gas to from 7 to 9 of air produces the best working effect. Its calorific value varies in different localities from 550 to 650 heat units per cubic foot. Ordinary oil illuminating gas varies somewhat in its constituents, and may average: Hydrogen, 39.5; marsh gas, 37.3; nitrogen, 8.2; heavy hydrocarbons, 6.6; carbonic oxide, 4.3; oxygen (free), 1.4; water vapor and impurities, 2.7; total, 100; and is equal to 617 heat units per cubic foot.

Producer Gas.

The constituents of producer gas vary largely in the different methods by which it is made; in fact, all of the following gases are made in producers, so called. The constituents of the low grade of this name are:

Carbonic oxide, CO	22.8 per cent.
Nitrogen, N	63.5 "
Carbonic acid, CO_2	3.6 "
Hydrogen, H	2.2 "
Marsh gas (methane), CH_4	7.4 "
Free oxygen, O	.5 "
	100.0 "

The average heating power of this variety of producer gas is about 111 heat units per cubic foot.

Another producer gas, called

Water Gas,

has an average composition of—

Carbonic oxide, CO	41 per cent.
Hydrogen, H	48 "
Carbonic acid, CO_2	6 "
Nitrogen, N	5 "
	100 "

and has an average calorific value of 291 heat units per cubic foot.

Semi-Water Gas,

or, as designated in England, *Dowson gas,* from the name of the inventor of a water gas-making plant, has the following average composition:

Hydrogen, H	18.73 per cent.
Marsh gas, (methane), CH_4	.31 "
Olefiant gas, C_2H_4	.31 "
Carbonic oxide, CO	25.07 "
Carbonic acid, CO_2	6.57 "
Oxygen, O	.03 "
Nitrogen, N	48.98 "
	100.00 "

It has a calorific value of about 150 heat units per cubic foot.

PETROLEUM PRODUCTS USED IN EXPLOSIVE ENGINES.

The principal products derived from crude petroleum for power purposes may commercially come under the names of gasoline, naphtha (three grades, B, C, and A), kerosene, gas oil, and crude oil.

The first distillate: Rhigoline, boiling at 113° F., specific gravity 0.59 to 0.60; chimogene, boiling at from 122° to 138° F., specific gravity 0.625; gasoline, boiling at from 140° to 158° F., specific gravity 0.636 to 0.657; naphtha " C " (by some also called benzin), boiling from 160° to 216° F., specific gravity

0.66 to 0.70; naphtha "B" (ligroine), boiling at from 200° to 240° F., specific gravity 0.71 to 0.74.; naphtha "A" (putzoel), boiling at from 250° to 300° F.

The commercial gasoline of the American trade is a combination of the above fractional distillates, boiling at from 125° to 200° F., specific gravity 0.63 to 0.74.

Kerosene, boiling at from 300° to 500° F., specific gravity 0.76 to 0.80.

Gas oil, boiling at above 500° F., specific gravity above 0.80.

Crude petroleum, boiling uncertain from its mixed constituents, specific gravity about 0.80.

The vapor of commercial gasoline at 60° F. is equal to 130 volumes of the liquid, sustains a water pressure of from 6 to 8 inches, and will maintain a working pressure of 2 inches, or equal to any gas service when the temperature is maintained at 60° F., and with an evaporating surface equal to $5\frac{1}{4}$ square feet per required horse-power, using proportions of 6 volumes of air to 1 volume of gasoline vapor.

Commercial kerosene requires a temperature of 95° F. to maintain a vapor pressure of from $\frac{1}{4}$ to $\frac{1}{2}$-inch water pressure, requiring a much larger evaporating surface than for gasoline. It may be vaporized by heat from the exhaust, and is so used in several types of oil engines.

TABLE IV.—PERCENTAGE, SPECIFIC GRAVITY, AND FLASHING POINT OF THE PRODUCTS OF PETROLEUM.

Products.	Per cent. of each.	Specific gravity.	Flashing point, Fah.
Rhigolene and chimogene	Trace.		
Gasoline	.02	0.650	10°
Benzine naphtha	.10	0.700	14
Kerosene, light	.10	0.730	50
Kerosene, medium	.35	0.800	150
Kerosene, heavy	.10	0.890	270
Lubricating oil	.10	0.905	315
Cylinder oil	.05	0.915	360
Vaseline	.02	0.925	
Residuum and loss	.16		
	1.00		

Crude petroleum and kerosene are available also by injection in a class of oil engines of the Hornsby-Akroyd type, in which the oil can be so atomized and vaporized as to make its entire volume available as an explosive combustible, in order that the accumulation of refuse shall be at a minimum. Crude oil is also used in the "Best" oil-vapor engine

CARBURETTERS.

The use of the vapor of gasoline, naphtha, and petroleum oil for operating internal-combustion engines is increasing to a

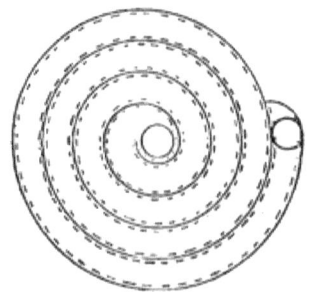

FIG. 14.—THE CIRCULAR CARBURETTER, PLAN.

vast extent in all parts of the civilized world, and will be no doubt the cheapest medium for generating power so long as petroleum and its products are at the present low price. In

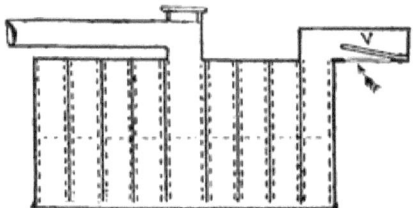

FIG. 15.—THE CIRCULAR CARBURETTER, SECTION.

gas-engine running, air saturated with the vapor of gasoline and naphtha is in general use, and when so used is produced by passing air through the liquid or over a surface largely extended by capillary attraction of the fluid by fibrous surfaces dipping into the fluid, by vaporizing the fluid by means of the heat of the exhaust, and by injecting the fluid in small portions into the air-inlet chamber or under its valve, and directly into the clearance space of the cylinder.

In Figs. 14 and 15 is illustrated a form of carburetter,

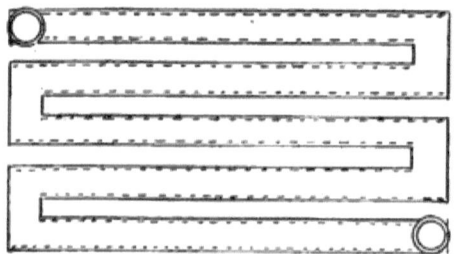

FIG. 16.—PLAN OF VENTILATING CARBURETTER.

made by the writer many years since, for carburetting air and low-grade illuminating gas.

This carburetter may be made of heavy tinplate. The spiral partition, made of tinplate, is perforated with sufficient small holes at top and bottom to fasten strips of cotton or woollen flannel on both sides of the spiral plate by stitching with coarse

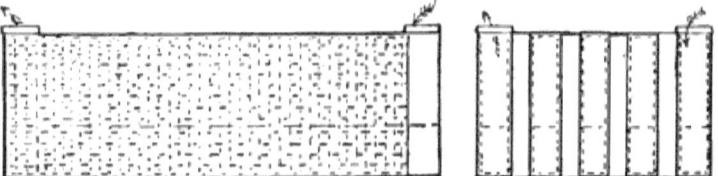

FIG. 17.—SECTIONS OF VENTILATING CARBURETTER.

thread and needle. The spiral plate should extend so as to nearly touch the bottom of the tank; the bottom is to be soldered on last. The valve V, for the purpose of preventing the escape of the vapor when the carburetter is not in use, may be made as light as possible, of tin plate or brass, and faced with soft leather wet with glycerin or a composition of glycerin and glue jelly, which always keeps soft and is not injured by the gasoline or its vapor. By this arrangement many square feet of surface may be obtained in a small space and perfect uniformity of saturation insured. As the enclosed walls of this form become very cold by long-continued use, an improvement was made by

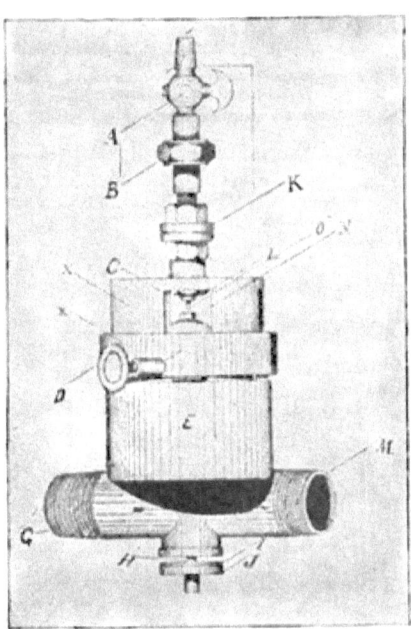

FIG. 18.—UNION AND GLOBE ENGINE VAPORIZER.

making each division wall with an outside surface, so that there was a natural down-draught of air on the outside of the entire evaporating surface of the carburetter. In Figs. 16 and 17 are shown the plan and sections.

In this form the air spaces prevent excessive cold by a circulation of air downward against the cooling surface of the walls—the whole interior vertical walls being lined with cloth fastened to a wire frame made to fit each section and pushed into place before the ends of the sections are soldered on.

Very good carburetters have been made by a long cast-iron

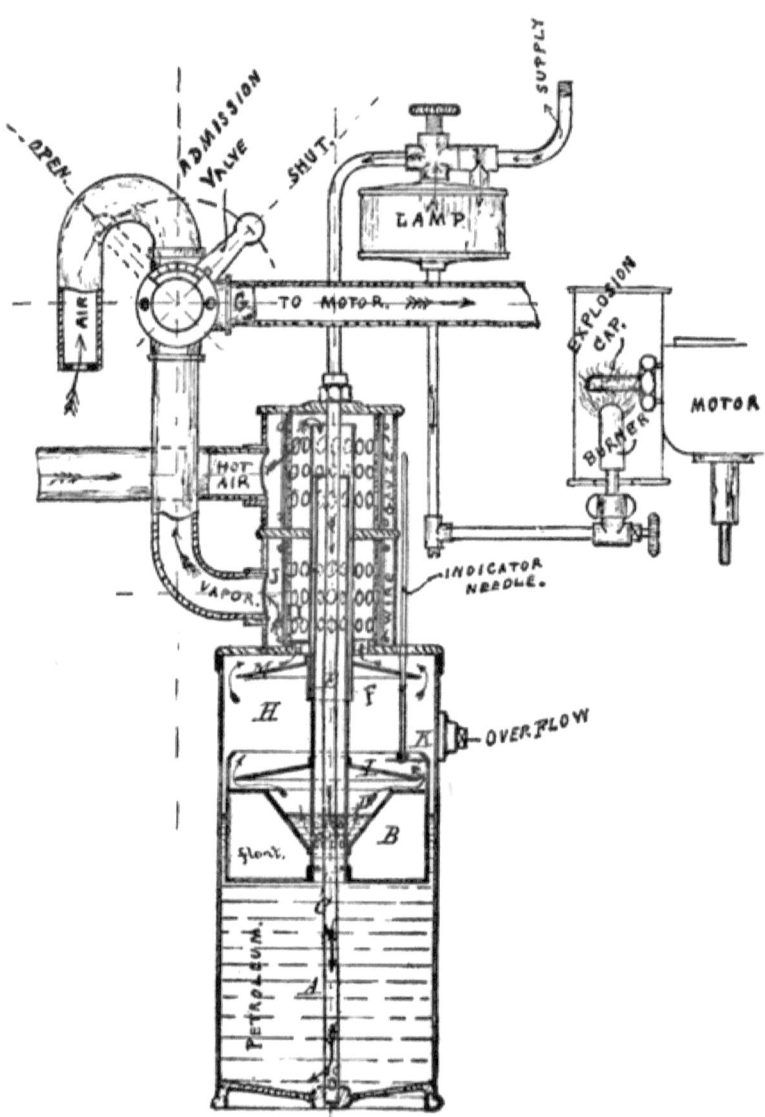

FIG. 19.—THE DAIMLER CARBURETTER.

box with a cover bolted on with a packing of glue and glycerin jelly on felt or asbestos packing, in which a frame of wire-work and cloth or yarn is made to give the desired evaporating surface.

For any carburetter of the forms here described, the depth should be limited to 8 inches, as the capillarity of the fibrous material is of little or no value at a greater height than 6 inches above the fluid, which should not be charged above 3 inches in depth for best effect.

In Fig. 18 is represented a vaporizer used by the Globe Gas Engine Company of Philadelphia. It consists of a metal body E, inside of which is a ball-shaped valve N, seated on the end of a tube with its spindle extending below the air pipe and attached to a disc at J for regulating the lift of the air and gasoline valve; O is spindle of gasoline valve. The gasoline tank is so placed as to flow the liquid to the vaporizer. The air is heated by passing through a jacket on the exhaust pipe.

Fig. 19 represents a sectional view of the Daimler carburetter. The incoming air is heated by passing through a jacket on the exhaust pipe, and charged to saturation with vapor in the carburetter, the saturated air charge being regulated by a three-way cock, which allows a further dilution with air for the explosive mixture.

The gasoline supply is made through the small central tube to the bottom of the carburetter, which insures a uniform density in the fuel. The float B by its weight keeps a constant level in the conical cup D, where evaporation takes place. The float and its guide-pipe move down as the gasoline is used. The hot air passes down through the guide-tube and out through the perforation beneath the fluid in the conical cup D, then over two diaphragms, and through the perforated screen and to the vapor tube. The perforated screen in both inlet and outlet chamber prevents the jerky motion of the air caused by the suction of the piston. The lettering in the cut fairly explains the ignition arrangement.

In Fig. 20 is represented the carburetter of the Gilbert & Barker Manufacturing Company, Springfield, Mass. It is made of wrought iron, has four divisions, in which perforated capillary partitions are set around each division or story of the carburetter, thus greatly enlarging the evaporating surface. The air enters the lower compartment, becomes saturated, and leaves the carburetter from the top. Provision is made for

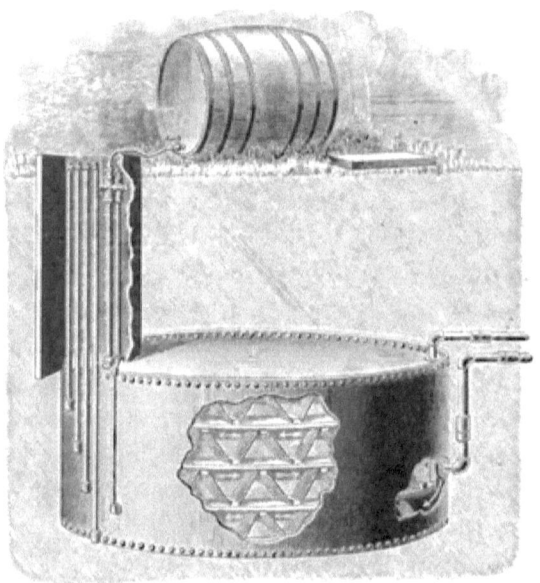

FIG. 20.—GILBERT & BARKER CARBURETTER.

pumping out any residue that may require removal when the carburetter is placed underground.

Many other forms of carburetter have been tried, without, however, securing better results than with those here described.

Saturated air with gasoline vapor has a heat value of about 200 heat units per cubic foot.

A claim has been made in France that by saturating part of the exhaust and by heating the gasoline, also by the exhaust, a concentrated vapor was produced, which, used with the air, produced a power value of $\frac{2}{100}$ of a gallon of gasoline per horse-power per hour. We await its confirmation. There is no doubt that greater economics are in progress in the operation of gasoline and oil engines; but the use of part of the products of combustion from the exhaust tends to lessen its value, if it has a value above its use as a part of the contents of the clearance space now in use in engines of the compression class.

The evaporation of gasoline of .74 specific gravity at a temperature of 60° F. varies somewhat from the form of its elementary constituents; so that an average of 1,173 grains per square foot of saturated surface per hour in the open air may be assumed as the basis for carburetting surface.

When evaporated in a closed vessel, as a carburetter, the vapor may start at about 1,000 grains per square foot of surface per hour; but if the area of evaporating surface is so extended that little or no tension or pressure is produced by its evaporation, due to the draught upon it by the motor, and the temperature of the gasoline is kept near to 60° F., the evaporation may be relied on at about 800 grains per square foot per hour.

This gives a basis for computing the area of carburetted surface at any assumed consumption of gasoline per horsepower per hour. For example, gasoline weighing 6 lbs. per gallon, with an assumed requirement of $\frac{1}{16}$ of a gallon per horse-power per hour, and an evaporation of 800 grains per hour per square foot, will require $\frac{\frac{6}{16} \times 7000}{800} = 5\frac{1}{4}$ square feet of evaporating surface in the carburetter per horse-power.

CHAPTER VIII.

CYLINDER CAPACITY OF GAS AND GASOLINE ENGINES.

The cylinder volume of gas and gasoline engines seems to be as variable with the different builders as it is with steam engines in its relation to the indicated power.

The proportion of diameter to stroke varies from equal measures up to 38 per cent. greater stroke than the measure of the cylinder diameter. The extreme volumes of cylinder capacity (measured by the stroke) varies from 28 to 56 cubic inches for a 1 H.P. engine and from 48 to 98 cubic inches for a 2 H.P. engine; for a 3 H.P. engine from 77 to 142 cubic inches, while for a 6 H. P. engine it ranges from 182 to 385 cubic inches. This disparity in sizes for equal indicated power may be caused by the different kinds of gas and its air mixtures under which the trials for indicated power may have been made, or it may be partly due to relative clearance and facility for exploding the charge at some fixed time.

It may be readily seen from inspection of the heat value of different kinds of gas—varying as they do from about 950 heat units per cubic foot for the highest illuminating gas to from 185 to 66 heat units in the different qualities of producer gas—that large variations in effective power will result from a given sized cylinder. It will also be plainly seen that with the extreme dilution of producer gas with the neutral elements that produce no heat effect, that no combination with air that also contains 80 per cent. of non-combustible element can produce even a modicum of power in the same sized cylinder as is used for a high-power gas.

In view of this it seems necessary to build explosive engines with cylinder capacities due to the heat unit power of the com-

CYLINDER CAPACITY.

bustible intended to be used, as well as to the method of its application.

In the following tables are given the indicated and actual power, revolutions, and size of cylinder and stroke of various styles of gas engines for comparison:

THE SINTZ.				THE ATKINSON CYCLE.			
Horse-power.	Revolutions per minute.	Diameter of cylinder. Inch.	Stroke. Inch.	Horse-power.	Revolutions per minute.	Diameter of cylinder. Inch.	Stroke. Inch.
1........	425	$3\frac{1}{2}$	$3\frac{1}{2}$	2.......	180	$4\frac{3}{4}$	$4\frac{3}{4}$
2........	400	4	4	3.......	180	$5\frac{3}{4}$	$5\frac{1}{2}$
3........	375	$4\frac{3}{4}$	5	5.......	160	$6\frac{7}{16}$	$8\frac{1}{4}$
4........	350	5	6	7.......	150	$7\frac{1}{2}$	$8\frac{1}{2}$
6........	300	$5\frac{3}{4}$	6	9.......	150	$8\frac{1}{4}$	9
8........	270	$6\frac{1}{2}$	7	12.......	140	$9\frac{1}{2}$	$11\frac{3}{16}$
10........	250	8	8	16.......	130	10	$11\frac{1}{2}$
15........	225	9	9	20.......	120	12	$12\frac{1}{2}$

THE NASH.				PACIFIC.			
Actual horse-power.	Revolutions per minute.	Diameter of cylinder. Inch.	Stroke. Inch.	Actual horse-power.	Revolutions per minute.	Diameter of cylinder. Inch.	Stroke. Inch.
$\frac{1}{4}$........	350	3	4	$1\frac{1}{2}$.......	250	$4\frac{3}{4}$	6
$\frac{1}{2}$........	350	$3\frac{1}{2}$	4	$4\frac{1}{2}$.......	225	$6\frac{1}{2}$	9
1........	325	4	$4\frac{1}{2}$	6........	200	7	10
2........	300	5	5				
3........	300						
4........	300						
5........	280						

LAWSON ENGINE.				STAR.			
Actual horse-power.	Revolutions per minute.	Diameter of cylinder. Inch.	Stroke. Inch.	Actual horse-power.	Revolutions per minute.	Diameter of cylinder. Inch.	Stroke. Inch.
1........	180	$4\frac{1}{2}$	8	2.......	250	$4\frac{1}{2}$	6
2........	160	5	10	3.......	240	5	6
4........	160	$6\frac{1}{2}$	12	4.......	220	$5\frac{1}{2}$	10
6........	160	$7\frac{1}{2}$	14	6.......	220	$6\frac{1}{2}$	12
				8.......	180	7	13
				10.......	180	8	14

RATING OF SOME ENGLISH ENGINES.

Indicated horse-power.	Revolutions.	Diameter. Inch.	Stroke. Inch.	Name.
9	164	6	16	Crossley.
9	164	8	16	"
14	200	7	15	"
16	160	$11\frac{1}{2}$	20	Burt's Otto.
18	180	$9\frac{1}{4}$	16	" "
19	160	$9\frac{1}{2}$	18	Crossley.
20	184	$9\frac{3}{4}$	17	Stockport.
20	164	12	18	Wells.
24	180	10	18	Barker's Otto.
30	170	12	20	" "
33	210	17	$21\frac{1}{4}$	Crossley.
40	160	18	24	Tangye.

The apparent discrepancies in the above tables of cylinder capacities, as to their size when compared with their indicated power, are not really so great as may be noticed at first inspection; for the mean pressure varies very much with the various fuels, as well also from the relative variation of the proportion between the volume of the combustion chamber and the volume swept by the piston. The difference in speed between the various engines noted also complicates the direct comparison for cylinder capacities.

The whole subject of size and weight of explosive engines for stated powers appears to be still in the experimental stage, which by continued experiment and experience may be brought into an approximate uniformity in practice.

MUFFLERS ON GAS ENGINES.

The method of muffling the sound of the exhaust, as well also the sound or clack of the valves, was a puzzling problem to the early builders of gas engines. The matter has finally sifted down to a plain cast-iron box of from 1 to 3 cubic feet capacity, set near the engine, and into which the exhaust pipe is connected, and continued by a separate connection to the outside of a building.

Connection of the exhaust with a chimney should not be made under any circumstances, as there are unknown elements of explosion liable to be accumulated in the line of the exhaust that might do damage to a chimney; and for the same reason the muffler-box should be made strong enough for a pressure equal to the explosive power of the gas and air mixture, or say 175 lbs. per square inch. This insures safety from any explosion that may accidentally occur in the exhaust by missed explosions in the cylinder, or otherwise.

The muffler pot is also a water-catch, in which part of the water vapor formed by the union of the hydrogen and oxygen is condensed. It should have a draw-off cock a few inches above the bottom, so that the muffler may always have a little water in the bottom, the water having been found to have a deadening effect on the exhaust.

A second muffler pot has been found to still further deaden the exhaust, and is preferable to throttling the exhaust by mufflers with perforated diaphragms.

In all cases an enlargement of the exhaust pipe from the muffler to the roof by one or two sizes larger than the engine exhaust, will modify the intensity of the exhaust at the roof, and often abate a nuisance.

CHAPTER IX.

GOVERNORS.

The regulation of the speed of explosive engines has an important bearing upon their usefulness and freedom from

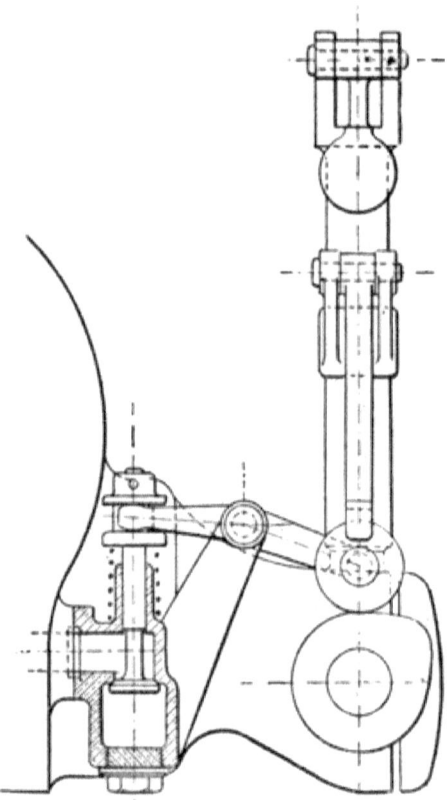

FIG. 21.—THE ROBEY GOVERNOR.

constant personal attention. By experience from trials during the few years of the growth of the new motor, much progress has been made in perfecting the details of this important adjunct of safety and uniformity in speed regulation through the action of a governor. There are four principal methods in use

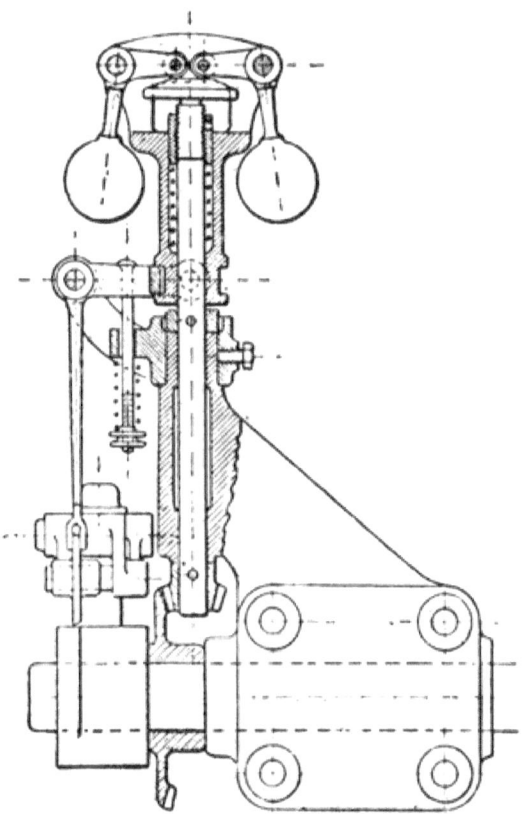

FIG. 21A.—THE ROBEY GOVERNOR.

for controlling the speed, viz.: (1) By graduating the supply of the hydrocarbon element; (2) by completely cutting off the supply during one or more revolutions of the crank; (3) by holding the exhaust valve open or closed during one or more strokes; (4) in electric ignition by arresting the operation of the sparking device.

To vary the quantity of the hydrocarbon by the action of

the governor is claimed to be the most economical as well as the most satisfactory method in use, if the variation in the work of the engine does not carry the charge beyond the limit of combustion; otherwise the second method seems to give the best results.

In Figs. 21 and 21A are two elevations of the centrifugal ball

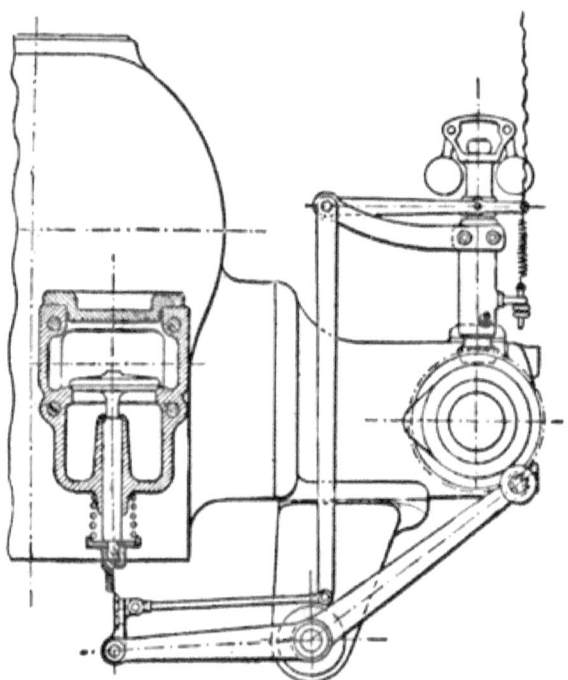

FIG. 22.—THE PICK-BLADE GOVERNOR.

governor, as used on the Robey and other engines in Europe and adopted with many variations on a number of American engines. In this type the bell-crank arm of the governor, by its centrifugal action, raises or depresses a yoke and sleeve which operates a bell-crank lever with a forked end astride a rotating disc which rides on the cam of the secondary shaft. The disc has a lateral motion on the end of the valve lever, so that the action of the governor rides the disc on to or off the cam, and thus makes a hit-or-miss stroke of the valve.

The centrifugal governor (Fig. 22) is another application of the hit-and-miss principle, by the use of a pick-blade operated

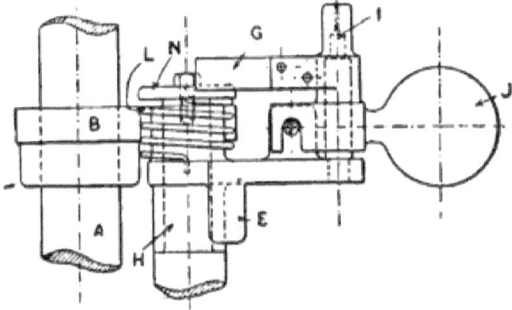

FIG. 23.—INERTIA GOVERNOR, PLAN.

from the governor by a balanced bell crank and connecting rod. The cut fully explains the detail of its construction and operation, by which an abnormal speed of the governor pulls the

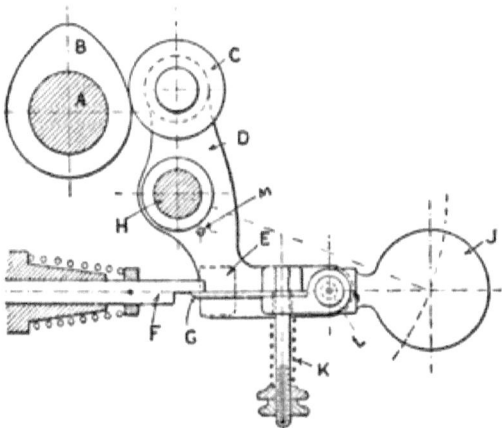

FIG. 24.—INERTIA GOVERNOR, ELEVATION.

pick blade away from the gas-valve spindle. In some forms graduated notches are made on the pick-blade or spindle-blade, so that in action the governor gives a varying charge within certain limits and a mischarge when the speed is beyond the limitation.

The inertia governor used on the Crossley engine in Eng-

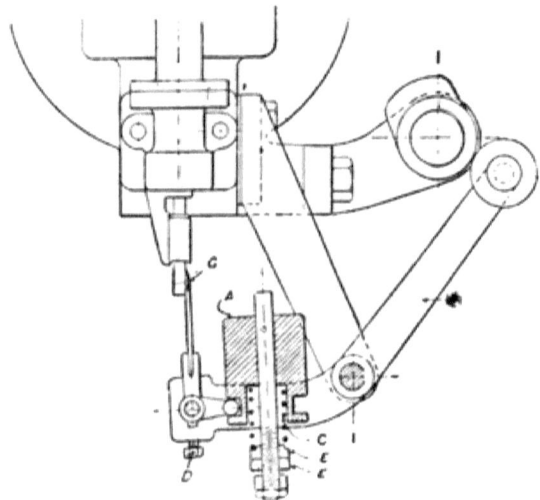

FIG. 25.—THE VIBRATING GOVERNOR, ELEVATION.

land, and with many modifications in use on American engines, is illustrated with plan and elevation in Figs. 23 and 24, in which A is the cam shaft, B cam, C roller, D lever, H lever

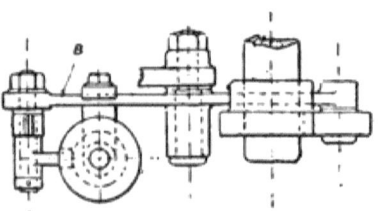

FIG. 26.—THE VIBRATING GOVERNOR, PLAN.

pin, L spring to hold the roller C to the cam, J the governor weight, K the adjusting spring, G the pick-blade, and F the valve stem.

In the action of this governor the initial line of motion of the ball J, in regard to its centre of motion H, is shown by the dotted curved line. By the sudden movement of its pivoted centre L, the ball is retarded in its motion by the regulating spring K, which tends to throw the pick-blade G off the shoulder of the valve F.

It will be readily seen that the inertia of the vibrating ball

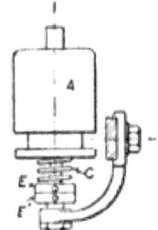

FIG. 27.—END VIEW, ELEVATION.

will vary as the speed of vibration, so that by carefully adjusting by the spring K, the action of the ball will vary the disengagement of the pick-blade to correspond with the over-speed of the engine, and make an entire miss at the limit of its variation. The air valve may also be operated by the spud E.

Another form of governor, involving the same principles of

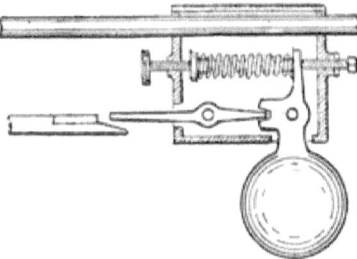

FIG. 28.—THE PENDULUM GOVERNOR.

inertia as the last one, is used on the Stockport engine in England, and is illustrated in Figs. 25, 26, and 27. It consists of a weight A, balanced on the vibrating arm B. A groove around the weight operates a bell crank, to which the pick-blade is attached. The balance spring is adjustable for regulating the position of the pick-blade and its contact with the valve spindle. By the variation in overcoming the inertia of the weight by the spring with different vibrating speeds in the lever, the disengagement of the pick-blade with the spindle-blade is varied or a mis-stroke made.

The pendulum governor (Fig. 28) is also an inertia governor in the principle on which it operates. It is attached to the exhaust-valve push-rod, and vibrates horizontally with the rod. The weight or ball has an extension or neck, with a pivoted eye, a yoke, and a vertical lug. The eye is pivoted in the box, and the yoke embraces the push-blade stem, which is also pivoted horizontally with the eye in the box or frame. The lug bears on an adjusting spring, which is set up by a screw so as to limit the swing of the ball to the normal speed of the engine, so that when the speed rises above the normal the inertia of the ball holds it back in its vibration and lifts the push-blade out of contact with the valve-stem.

In some engines the position of the ball is reversed, and it stands above the valve push-rod on a finger and is made adjustable in its length of oscillation by its distance from the fulcrum.

Several modifications of the governors here described are in use, devised on the principles of inertia as illustrated in Figs. 24, 25, and 28.

CHAPTER X.

IGNITERS AND EXPLODERS.

THE devices for firing the charge in gas, gasoline, and oil engines may be divided into four types, with as many varia-

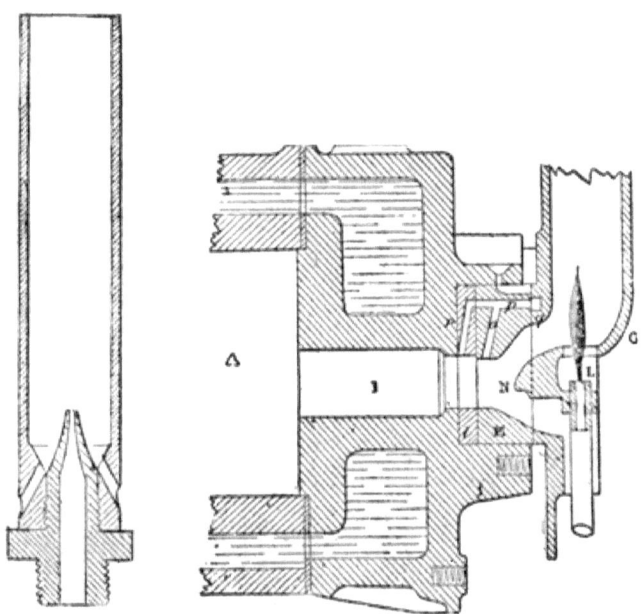

FIG. 29.—THE BUNSEN BURNER.

FIG. 30.—THE OTTO IGNITION SLIDE-VALVE.

tions in the form of each type as may suit the requirements of construction or the fancy of designers.

The simplest arrangement is probably the direct-flame contact of a gas-burner in contact with the walls of the cylinder, with a hole through the cylinder wall that is uncovered at the proper moment for ignition by the movement of the piston, as in the earlier two-cycle non-compression engines — the in-

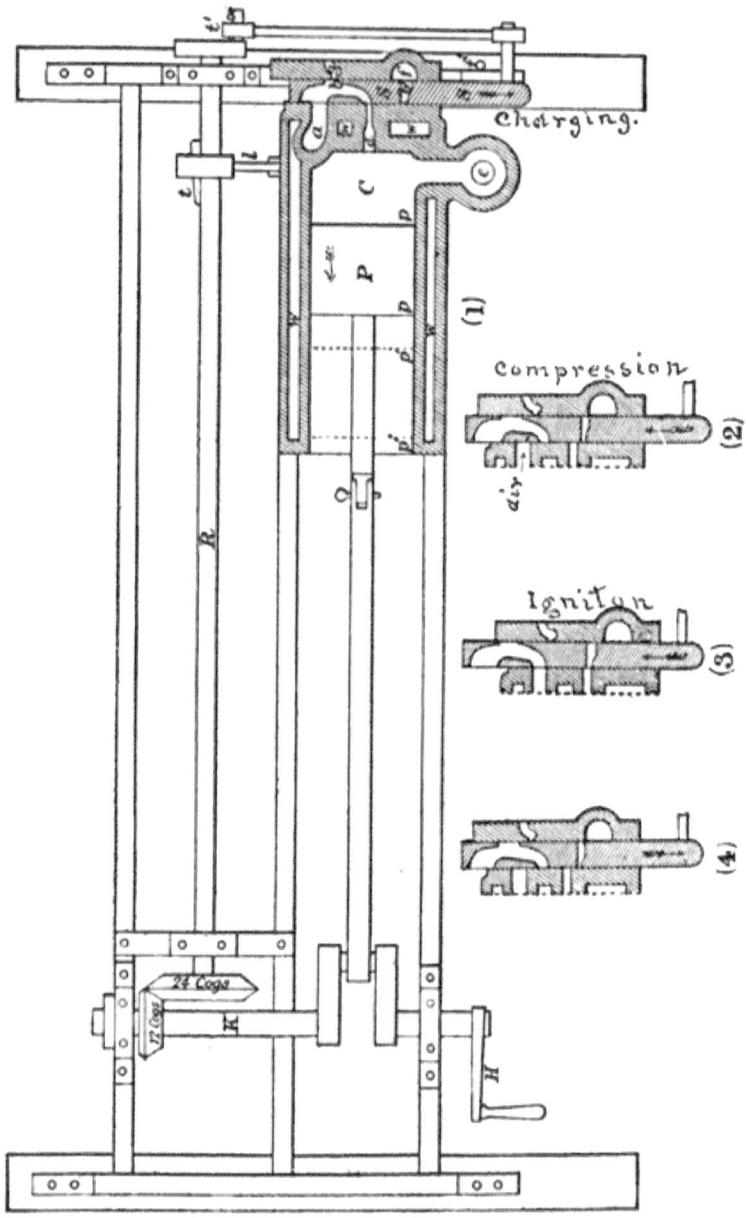

FIG. 31.—OPERATION OF THE OTTO IGNITION SLIDE-VALVE.

draught of the flame and explosion taking place at the point in the stroke at which the charge of gas and air mixture is completed. This igniter may be in the form of a partially aërated gas or vapor mixture, flowing through a tube constructed like a Bunsen burner, as shown in Fig. 29, the burner being set with its mouth just below the igniting port in the cylinder, with an outside guard tube to keep the flame steady; or a large flame may be used in contact with the port, as shown in the illustration of the economic gas engine, further on.

This form of igniter is also used on compression engines of the four-cycle type, with slide-valves enclosing ignition chambers, notably on European and American engines of the Otto slide-valve type.

Fig. 30 shows a section of a cylinder head with position of flame, guard chimney, and slide-valve at the moment of ignition.

Fig. 31 is a sectional view of the ports in the slide-valve and cylinder head of an Otto slide-valve engine, showing the position of the ports at different points in the stroke. No. 1, cylinder charging with air and gas, in which a is the air port, g the gas port, b the back port in the slide s, and b' the ignition port. No. 2, position of the slide during the return or compression stroke. No. 3, movement of the ignition port from the flame to the cylinder port. No. 4, reversal of the slide movement during the pressure stroke.

Fig. 32 illustrates the piston igniter as used on some of the Nash engines, where e is the gas jet, d opening through the valve shell, g the passage into the ignition chamber.

This igniter is based upon a new principle. The igniting jet of combustible mixture is caused to rotate in the circular chamber r in the piston, into which it enters through a passage tangentially placed. This forms a vortex of flame, which is positive in its action and simple. The piston valve is made of steel, and is hardened and ground to size. It moves in a reamed hole in the case, being so loosely fitted as to drop of its own weight, and yet making a gas-tight joint. Since the valve is perfectly balanced as to gas pressure, it moves without fric-

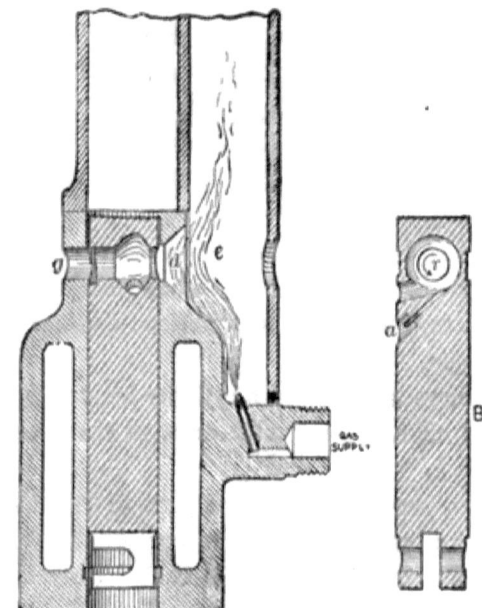

FIG. 32.—IGNITING VALVE.

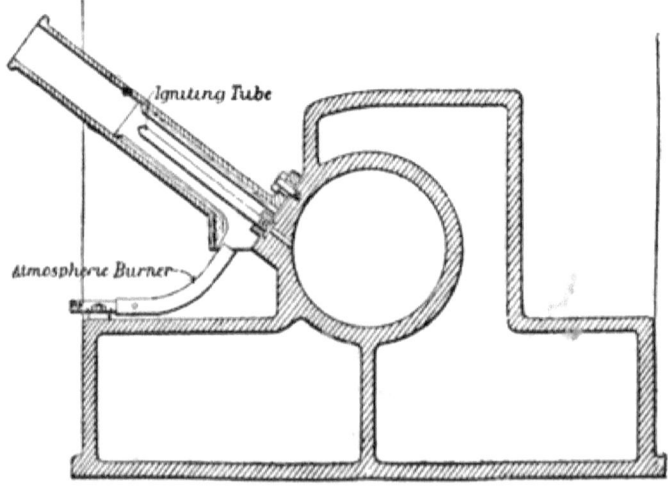

FIG. 33.—THE TUBE IGNITER.

tion, and therefore requires a very small quantity of oil—just sufficient to prevent it becoming dry. The valve is made long, and the lower part has a bearing in that part of the case kept cool by a water-jacket. As oil is only applied to the lower end,

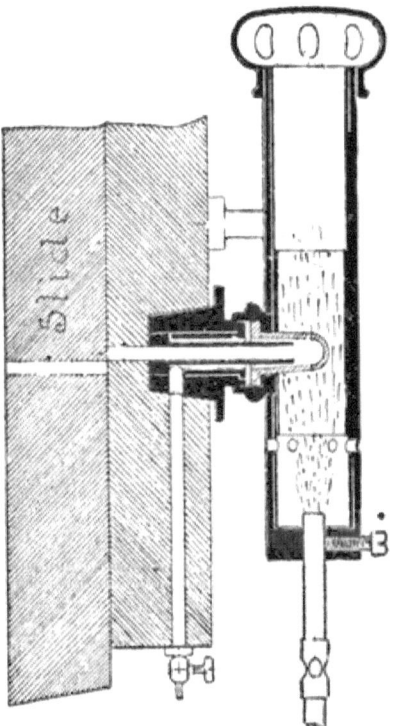

FIG. 34.—SLIDE IGNITER.

very little can work up to the hot end where the igniter is heated; hence the formation of gummy oil is prevented, and the valve seldom needs cleaning. In actual use it has been found that the case and upper end of the valve never come into metallic contact, as, on account of the looseness of fit at that point, a scale of hard carbon is formed over the surface of each, which protects them from abrasion. The valve is positively operated by an eccentric on the shaft.

The tube igniter, as shown in Fig. 33, has taken a wide range of usefulness and is well adapted to compression engines. As originally made, there is a deviation in the time of ignition from the uncertain condition of the explosive mixture and va-

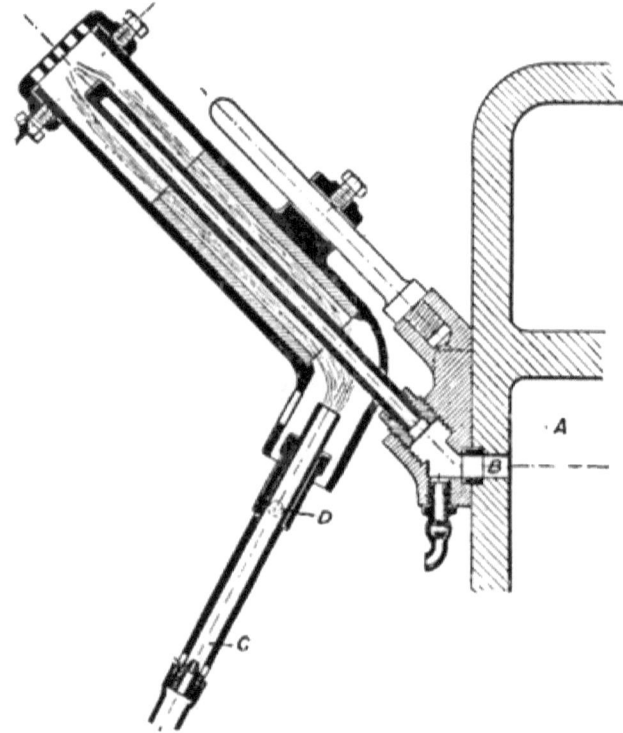

FIG. 35.—TUBE IGNITER.

riable heat of the tube. The adjustment of the length of the tube and position of the heating flame, so that ignition will take place at the maximum compression or end of the compression stroke, is a somewhat delicate matter, but has been found by experiment for the different designs of gas engines.

The degree of compression to just carry the fresh gas and air mixture to meet the firing temperature of the tube by pushing the products of the previous combustion before it, together with the adjustment of the Bunsen jet to a proper position in regard to the length of the tube, is a puzzling problem that has to be worked out experimentally for each style of engine.

In Fig. 34 is shown the form of slide igniters as used on

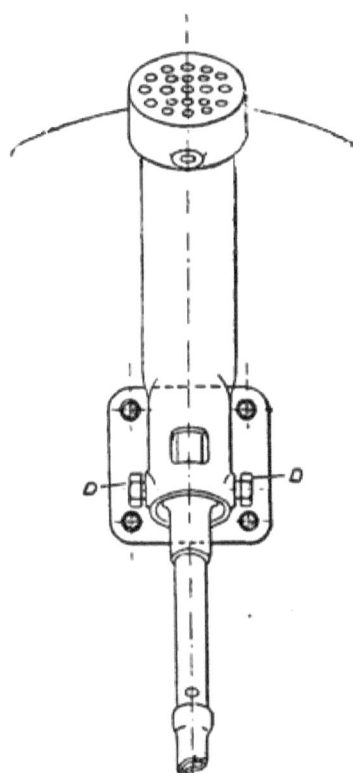

FIG. 36.—FRONT VIEW.

European engines using both tube and slide. This form acts as a time igniter, which regulates the time of ignition by the movement of the slide-valve or inlet piston, which opens communication with the hot tube through the inner tube by compression—the small vent tube and cock allowing of a free blowout of the igniting tube when accumulation of soot takes place.

In this plan the ignition tube is short, and may be made of platinum or porcelain.

The hot-tube igniter (Figs. 35 and 36) shows two views of an ignition tube used on the Robey engines, which is adjustable for the position of the igniting surface of the tube as well as for the position of the Bunsen burner, A being the combustion chamber, B the igniter passage, C the Bunsen burner pivoted to the chimney frame at D, which allows the burner to be tilted slightly to regulate the distribution of the flame around the tube.

The set-screw in the chimney socket allows of a ready adjustment of the position of the chimney and burner for the time of ignition.

ELECTRIC IGNITION.

Electric ignition involves three principles of action, viz.: the secondary-current spark derived from a battery and induction coil, with the spark transmitted between two platinum electrodes by a contact-breaker, operated by the valve shaft on the outside of the cylinder. The battery may be primary or storage, with the circuit-breaker connected between the battery and induction coil, as in Fig. 37, which represents the electric igniter used on the Priestman engine. The only difficulty with the perfect action of this form of igniter is the necessity of cleaning the insulation surface of the plug carrying the electrodes. The insulators are two porcelain tubes, set in a brass or iron screw-plug and projecting on the end toward the piston, in order to carry the spark nearest to the fresh inlet mixture of air and vapor, as well also to increase the insulating surface and allow of easy cleaning by unscrewing the plug and wiping the porcelain surfaces, which become occasionally fouled with a carbon deposit which short-circuits the current and prevents a spark.

In some electric igniters using stationary electrodes, the in-

IGNITERS AND EXPLODERS.

duction or Ruhmkorff coil is used, which produces a more voluminous spark and which is claimed to be more reliable in its

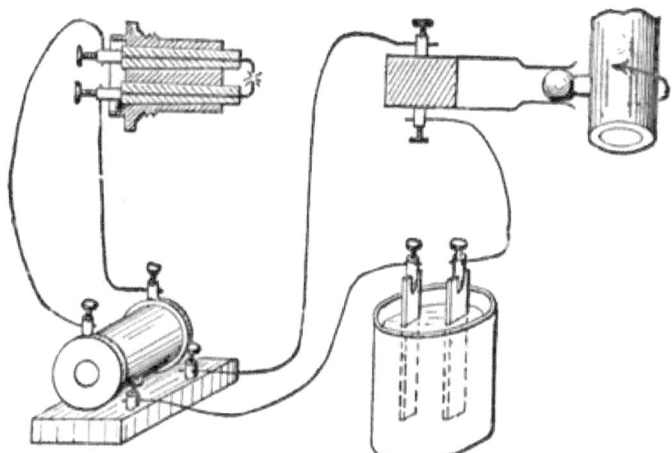

FIG. 37.—ELECTRIC IGNITER.

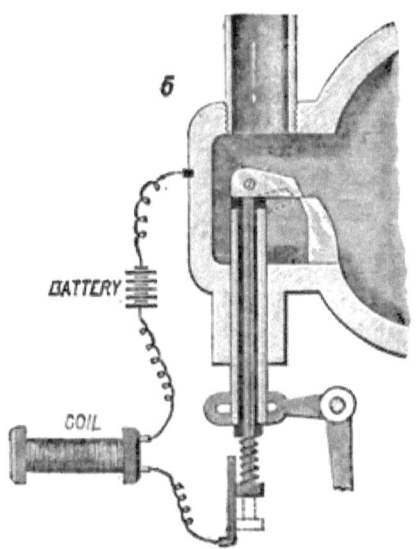

FIG. 38.—RECIPROCATING ROD SPARK-BREAK.

action and less liable to short-circuit by slight fouling of the insulator.

Several forms of internal circuit-breakers have been devised, in which Fig. 38 represents a reciprocating rod which may be operated by a connecting rod with a cam. The insulation is made within a sliding tube, which allows of considerable motion in order to allow the contact piece to slip off suddenly from the stud which is fixed in the cylinder head.

In Fig. 39 is represented a similar device, in which the insulated rod rotates by an outside gear driven from the valve

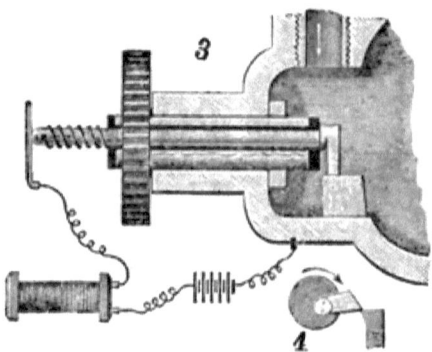

FIG. 39.—ROTATING SPARK-BRAKE.

shaft. The rotating spindle carries the insulated rod and break-piece eccentrically, so that its contact and break can be accurately regulated by rotating the position of the teeth of the gears.

The sparking coil used with this form of igniter is shown in Fig. 40. It consists of a bundle of iron wire, insulated and wrapped with insulated copper wire. It is a simpler device than the double or Ruhmkorff coil, but will not project a strong spark or at a great distance between the electrodes, as may be obtained from a Ruhmkorff coil—the breaking device being necessary in either case.

In Fig. 41 is represented the Pennington double igniter, in which the breaker is a loop piece attached to the end of the piston. The contact finger swings on a joint with a spring that keeps it in a straight line with the insulated rod. As the

IGNITERS AND EXPLODERS. 75

piston nears the end of its stroke, the loop pushes the finger over and breaks the contact at the end of the stroke; and as the piston recedes, the finger, having sprung back in line with the insulated rod, is caught by the loop, and a second break spark

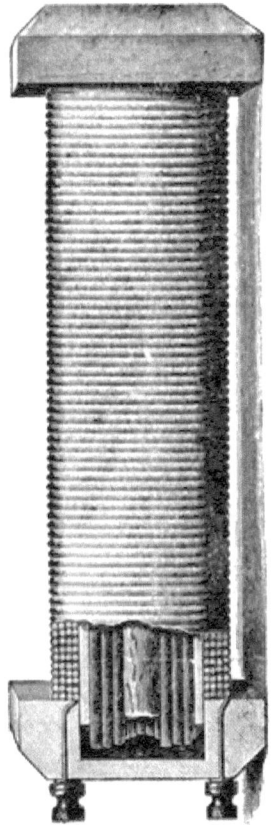
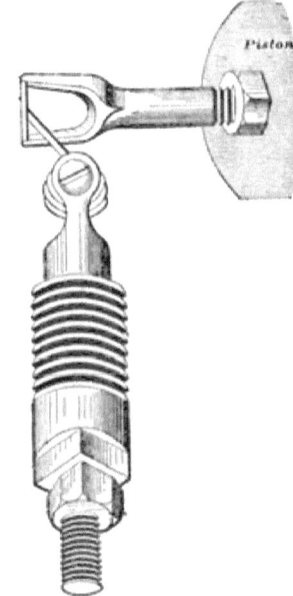

FIG. 40.—SPARKING COIL.   FIG. 41.—THE DOUBLE SPARK DEVICE.

takes place. The time of sparking can be varied by the length of the finger and by adjusting the position of the insulated plunger.

Ignition by direct current from a small dynamo with a current-breaker operated by the cam shaft is in favor with many gas-engine builders.

A current-breaker used on the Priestman engine is shown in Fig. 42, where an arm kept in position by a spring or weighted lever is made to touch a spud revolving on the sec-

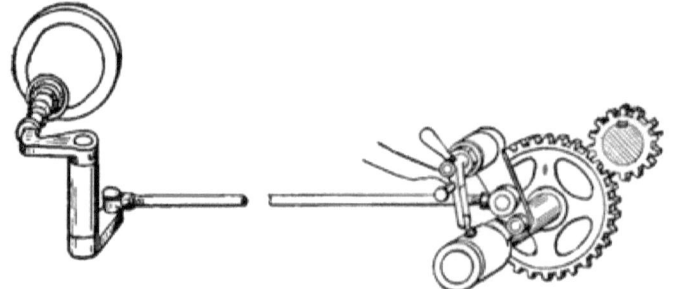

FIG. 42.—THE CURRENT-BREAKER.

ondary shaft. A movable sleeve on the shaft is set back or forward for time adjustment of the contact break.

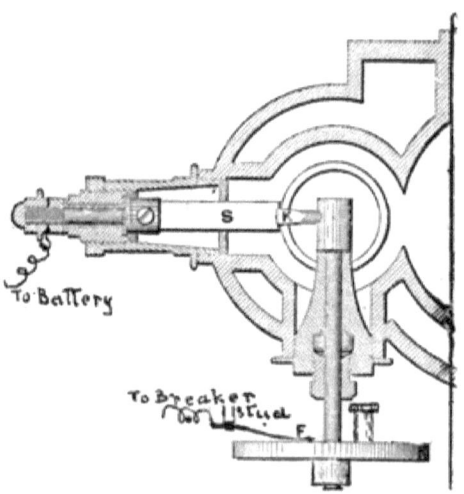

FIG. 43.—ROCKING SHAFT SPARKER.

Fig. 43 represents the sparking device used by the Union Gas Engine Company of San Francisco, and consists of a rocking shaft carrying a flattened pin, K, on the end inside of the

IGNITERS AND EXPLODERS. 77

firing chamber, which by its rocking motion is brought in contact with an insulated spring, S. The spring-contact piece, bearing against and rubbing the rocking pin, secures perfect freedom of current circuit while in contact.

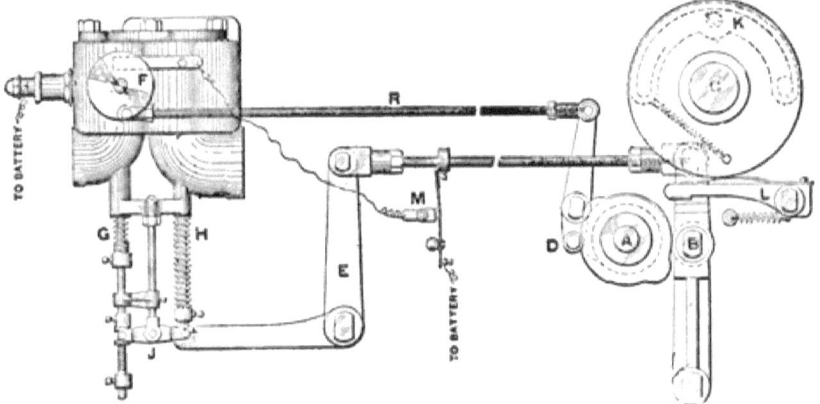

FIG. 44.—THE OPERATING DEVICE.

The operating device is shown in Fig. 44, where the push rod R, connecting with an arm moved by a cam on the second-

FIG. 45.—THE PERMANENT FIELD GENERATOR.

ary shaft, is adjusted to make the break contact at the required moment; while the contact spring at M relieves the battery circuit during the time of three cycles.

Ignition from the current of a small dynamo attached to the engine and driven at the proper speed from the engine shaft is in successful use and does away with the care of a battery. This requires no induction coil, the spark being made directly through the break device and electrodes.

Fig. 45 represents a generator used on the Sumner gas and gasoline engines. The spark is produced by a plunger contact with the commutator operated from a cam on the secondary shaft.

IGNITING TIMING VALVES.

The value of an exact time of ignition for producing uniformity of speed in explosive engines is attested by the ex-

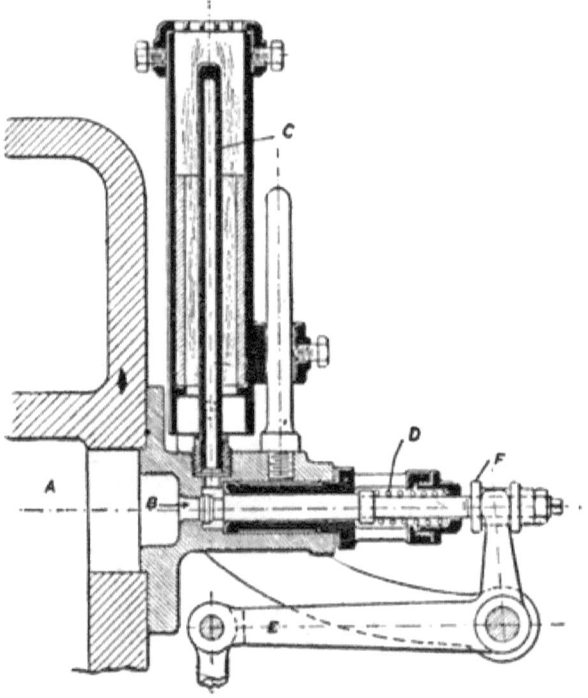

FIG. 46.—TIMING VALVE.

haustive experiments of years with the many devices made for the ordinary tube igniters, and the final recourse to electric ignition. A satisfactory result has been obtained in several designs for operating a valve at the mouth of the ignition tube that admits the compressed charge to the ignition tube at an exact point in the piston stroke.

In Fig. 46 is illustrated a timing valve used on the Robey

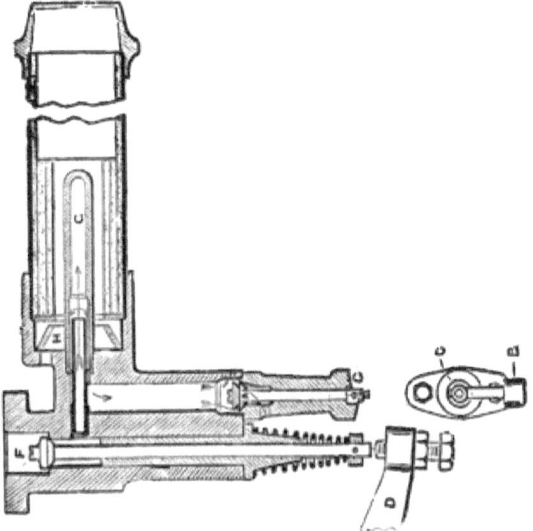

FIG. 47. TIMING VALVE AND STARTER.

engine, in which A is the combustion chamber; B the passage leading to the hot tube, a double-seated valve and spindle held to its front seat by the spring D; E a lever operated from the cam shaft; F adjusting spool with set nuts. In action the valve is opened at or about the end of the compression stroke and kept open during the exhaust stroke, thus clearing the ignition tube uniformly and insuring exact time of ignition.

In Fig. 47 is illustrated a combined timing-valve igniter and starter, as used on the Stockport engines. In this arrangement a double tube is used, with an annular space between the inner tube and the hot tube, through which the products of combustion may be blown out, followed by the explosive mixture, into the hot tube, by compressing the timing valve and the starting valve at the same moment. Referring to the cut, F is the timing valve, operated by the lever D; A the starting-valve, with its waste outlet at V; H is a mantle to draw the flame closer to the igniting tube.

There are many variations in form and attachments for timing valves in use in Europe and the United States. They are fast coming into favor for hot-tube igniters for the larger gas and gasoline engines.

CHAPTER XI.

CYLINDER LUBRICATION.

THE lubrication of cylinders of explosive motors is a matter of great importance, as the intensely hot gases in immediate contact with the lubricating oil, although the oil is in contact with a comparatively cool metallic surface, has an evaporative

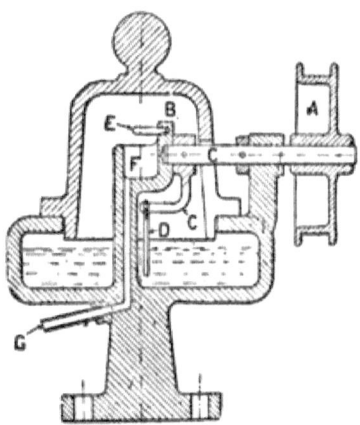

FIG. 48.—THE MECHANICAL LUBRICATOR.

effect, tending to thicken the oil into a gummy lining on the surface of the cylinder. To avoid this and keep a perfect lubrication, an oil that is adapted to this severe heat trial should be used and fed to the cylinder walls and piston in constant flow, and not too much or too little, but just enough so that the oil cannot be pushed into the combustion chamber in excess, so as to be blown through the exhaust valve to clog the passages with oily soot.

The sight feed and capillary drop-oil feeders have been perfected to such an extent in the United States that they are almost in universal use. Yet on some engines with revolving valve-cam shafts, the facility for obtaining easily the motion

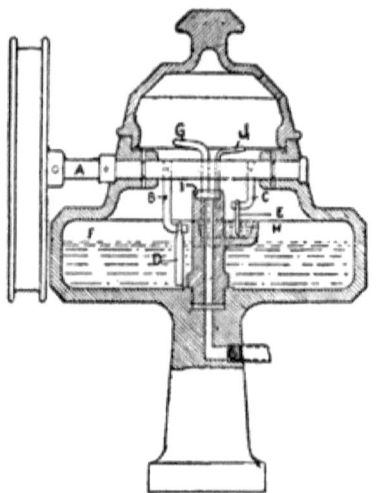

FIG. 49.—THE ROBEY OIL FEEDER, SECTION.

for a mechanical lubricator has kept this form in use on many engines.

In Fig. 48 is illustrated a mechanical lubricator used on the

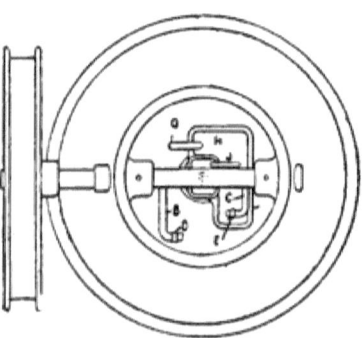

FIG. 50.—THE ROBEY OIL FEEDER, PLAN.

Crossley engines in England, and with some variations on other European and American engines. A small belt from the valve-cam shaft to the pulley A gives the required motion to the spindle and crank C C, to which is loosely attached a wire D, that dips into the oil and carries a minute portion to the wiper E, from which the oil drops into the passage to the cylinder.

In Figs. 49 and 50 is shown a section and plan of a lubricator used on the Robey engines, which is an improvement over the previous one, in that it has a small receptacle above the level of the main oil cistern, which is fed by a revolving shaft and crank arm with drop wire reaching to the bottom of the cistern and wiping the oil on a fixed wiper over the receptacle, from which a second crank arm and drop wire lifts the oil to the wiper that feeds the passage to the cylinder. By this arrangement the oil for the cylinder is drawn from a fixed level, and the feed is therefore perfectly uniform at any level of the oil in the cistern.

CHAPTER XII.

ON THE MANAGEMENT OF EXPLOSIVE MOTORS.

The drift of constructive practice in the United States seems generally to be in the line of simplicity and least number of parts, in order to conform to the needs of the people that have the care of such motive power. The explosive motor now appeals to no experience as an engineer for its care and running; yet it does seem to require some common sense as to cleanliness and the propriety of things that may assume a menacing or dangerous habit by neglect of some of the few points of attention required in persons having the charge of this rising prime mover. The ability to discover leakage of gas or oil vapors or the products of combustion in the pipe connections, through valves, or by a defective or worn piston; the thumping in journal boxes, looseness of pins and piston thump is easily acquired when a person assumes the care of an engine. The regulation of the explosive mixtures are fully explained in the instruction pamphlets and display sheets of the builders, and from the completeness of instructions furnished there seems nothing to fear in the first start of an explosive motor by any person of ordinary intelligence.

Cleanliness being of the first order, due attention should be given to the cleaning of the cylinder, valves, and exhaust pipe at stated intervals; in some motors at least once a month, in other motors several months may elapse without internal cleaning being necessary, apparently without detriment. But we apprehend that the quality of the fuel has much to do with the fouling of the combustion chamber and exhaust pipe, and therefore the quality of the fuel should be suggestive of the times indicated for internal cleaning. The outside surfaces should be wiped off before starting or at the close of work every day, especially where the location is in a room with working-people, as the odor of the lubricating oil is not agreeable when the oil is spread in excess over an engine.

In workshops or rooms where dust prevails it is most desirable to enclose the motor in a small room by itself, well ventilated from without, for motor cylinders are mostly open and gather dust on their oily surfaces, and dust in the ingoing air of combustion leaves grit and ashes in the cylinder. The oil for lubricating the cylinder should be of the best "cylinder oil" of the trade, and is sold by many dealers as "gas-engine cylinder oil." It is not so expensive as to preclude its use for all the moving parts of an explosive motor, although a poorer quality is in general use.

Automatic oil feeders are almost universally furnished with these engines, so that there should be very little waste of oil. In cleaning the internal parts from carbon and oil crust, no sharp scrapers should be used on any rubbing parts or the bearings of valves. If unable to remove the crust with a cloth and kerosene oil, a hardwood stick and oil will generally remove the incrustation down to the metal, while the valves, if not cut, only need rubbing on their seats with finely pulverized pumice or other polishing powder. Emery is not recommended, as valves often get too much grinding to their detriment by the use of this material.

In starting a motor it should always be turned over in its running direction, and when compression makes this difficult the relief valve (most motors have one) or the exhaust or air valve may be opened to clear the cylinder, if an overcharge of gas or a failure has been made at the first turn.

In most cases turning the fly-wheel two or three revolutions will clear and charge the cylinder under the usual conditions for starting. With most of the large motors a starting device is provided, which is described in Fig. 47, and in the special exhibit of the American explosive motors further on.

Some of the troubles to be met are severe explosions after several misfires, by which the cylinder may become overcharged with the combustible mixture. This is often caused by irregular work on the engine, and the consequent scavengering of the cylinder of the products of previous explosions, replacing with pure mixtures at the next charge. Again, by a misfire from failure in the igniter an explosive charge is intensified at the next ignition or exploded in the exhaust pipe. Other interruptions sometimes occur, such as the sticking of the exhaust valve open by gumming of the spindle or a weak spring. From this may also arise some of the back-firings in the muffler and exhaust pipe. All of these explosions taking place at irregular times may be attributed, first, to irregular work; second, to irregularity in the operation of the valve gear or igniter, and although not pleasant to the ear may not be considered dangerous, because the motors and all their parts subject to explosion are made equal in working strength to the greatest pressure made by such explosions.

With the compression usual in American motors, 40 to 50 lbs., the greatest force from misfire or back-fire explosives can scarcely reach 300 lbs. per square inch in the cylinders and 150 lbs. in the mufflers, unless, by a possible contraction of the exhaust pipe by carbon deposit, a muffler pot may have possibilities of rupture. In no case should an exhaust pipe be turned into a chimney. With gas engines the full power is sometimes not realized from insufficient gas supply. The gas bag is a good indicator of this condition, caused by a too small gas pipe or a small meter, by which a flabby appearance of the gas bag shows that the motor is drawing more than the pipe or meter can supply with a proper working pressure.

The muffler pots have been known to accumulate water in cold weather by condensation of the water vapor formed by the union of the hydrogen and oxygen of the gas and air, to such an extent as sometimes to cause fear in an attendant of a cracked cylinder and leakage of water in from the circulation.

The water should be drawn off occasionally from the muffler pot by a cock. Gas motors running with electric igniters sometimes do not start at first trial from the accumulation of air in the gas-pipe. Testing by a gas-burner or a second trial will show where the difficulty lies and its remedy. And finally, much caution should be observed in examining the interior of valve chambers and the electric exploders by taking off caps or plugs and using a light near them until assured that fuel inlets are closed and the motor has been turned over several times to clear it of all explosive mixture. The consequences of explosion from peepholes are obvious. Even when a motor has been idle for a time it should be opened with the above caution.

The adjustment of governors only require care and a careful study of the directions for operating the engines, as there are too many variations in the designs and methods of adjustment for definite instructions under this head. Much care is required in renewing the ignition tubes, especially after the spare tubes furnished with the engine have been all used. The same size gas-pipe and of the same length as the tubes furnished with the engine should be made and the end welded up or capped, so that they may contain the same volume as the original tubes. This caution will insure the uniform adjustment of the time of ignition by change of tubes; otherwise tinkering with the position of the Bunsen burner will not enable an attendant not experienced in regulating the time of ignition to regulate it with any degree of certainty. The regulation when once lost can be properly tested only by an indicator card.

With a timing valve and the amount of lead for the return fire from the tube being known, the adjustment of the timing-valve throw can be made from the position of the dead centre of the crank at the end of the forward stroke. The timing lead is the time that is required for the mixture to pass the valve and become compressed in the igniting tube and the flame to return to the combustion chamber, as measured on the circumference of the timing-valve cam.

Other than iron tubes are used, such as nickel, aluminum bronze, and porcelain, with satisfactory results. The porcelain tubes are made short and require a special fitting to adapt them to a chimney, or the chimney should be of special design (as shown in Fig. 34), for a cross impact of the flame of the Bunsen burner.

There are many points in the management of explosive motors that cannot be discussed in a general treatise, arising from the varied details of design, in which special reference to the methods of operating the valve gears of igniters and governors of each individual design is required. The special instructions furnished by builders are ample for the operation of their motors, and if carefully studied lead to success in their operation by any person of ordinary intelligence or tact in handling moving machinery.

CHAPTER XIII.

THE MEASUREMENT OF POWER.

THE methods of measuring power are of but two general forms or principles, although the individual machines or instruments for accomplishing the measurement are of many kinds and of a variety of construction.

The one form is especially adapted for the measurement of the available power of prime movers under the various conditions of the application of their elementary constituents, by the absorption of their whole output of power at the point of delivery and there record the value of its force and velocity. Its representative is the brake dynamometer, or Prony's brake, in the various details of construction that it has assumed as designed and applied to meet the views or fancies of mechanical engineers.

The second form is a marked departure from the structural form of the first, and with the principle in view of placing as little obstruction as possible to the transmission of power from the prime mover to the receiver of power, to measure the actual net or differential tension of a belt or gear, and with its velocity indicate the exact amount of power delivered to a line of shafting or a machine. These are called transmitting dynamometers in distinction from the absorption dynamometers of the Prony type. They are of two kinds, one with a dial and index pointer, by which the hand on the dial must be constantly watched and recorded for a length of time and a mean pressure obtained from the varying record. The other carries a self-marking register moved by clockwork, by which the actual pressure is a constant record for any desired time, or a full day's work, the only personal observation required being the speed of the pulley or belt or its average throughout the time or day.

In Fig. 51 we illustrate the first form, a simple absorption

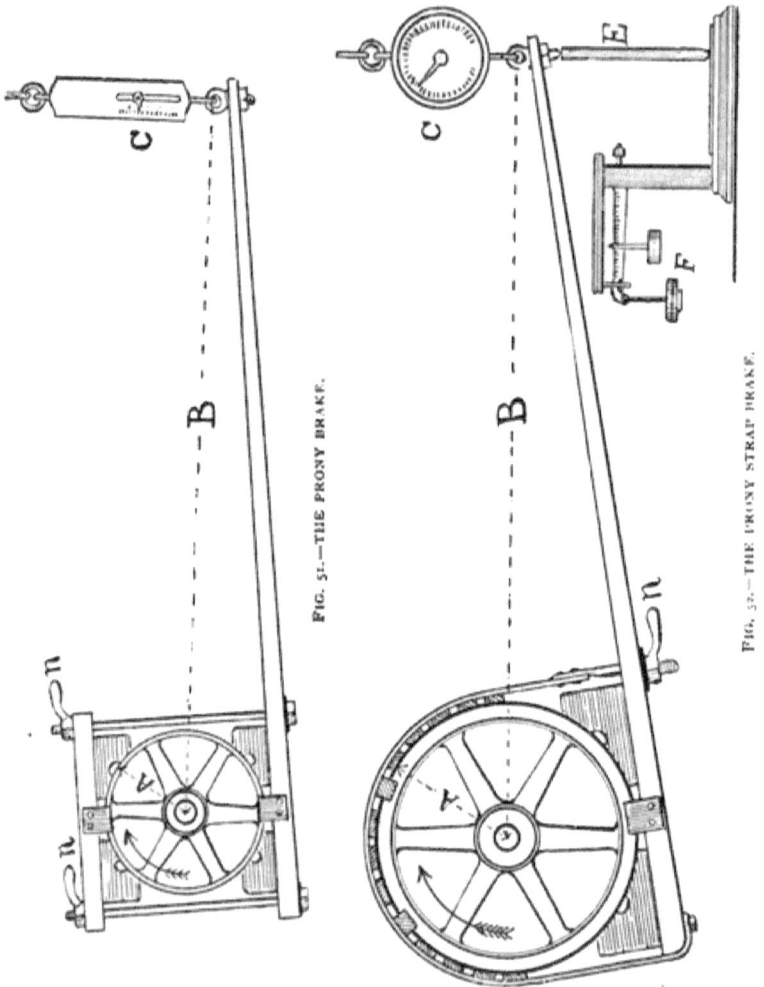

Fig. 51.—THE PRONY BRAKE.

Fig. 52.—THE PRONY STRAP BRAKE.

dynamometer or Prony's brake, named after its inventor, in which A is the radius of the pulley drum or shaft to which resistance may be applied; B the length of the lever from the centre of the shaft to the point of attachment of the spring scale or other means of measuring the tension of the lever; C a spring scale, which is preferable for light work within its range; and N N lever nuts for quick control of the pressure.

In Fig. 52 is presented a simple and inexpensive arrangemen of a power-absorbing brake for a large driving-pulley or finished fly-wheel, in which a belt is lined with blocks of wood spaced and fastened to the belt with screws or nails, a few of the blocks projecting over the edge with shoulders to prevent the belt from running off the pulley.

Spring scales may be purchased of the straight and dial pattern up to one or two hundred pounds capacity at reasonable figures, and are a source of satisfaction in showing the amount of vibration due to irregular pulsations of the motive element and crank motion. Where the measurement of power beyond the range of a spring balance is required, the use of a platform scale or any other weighing device may be made available. With a platform scale the light wooden strut, E, Fig. 52, may be adjusted to any length and vertically reaching from the platform to the horizon line, B, from the centre of the shaft; lanyards or any convenient means being used to keep the end of the lever from swaying.

Water from a squirt can is the best lubricant for this class of dynamometers, as it can be easily thrown upon the face of the pulley at the interstices of the blocks and lagging, and by its quick evaporation carries off the heat generated by friction. Soapy water has been used to good effect in preventing irregular pressure or stickiness of the friction surfaces.

It matters not in what direction the brake lever is placed to suit the convenience of observation, so long as the pull of the scale is made at right angles to the radial line from the shaft center. Its weight, as indicated on the scale, with the friction blocks or strap loosened in any position that it may be set, should be noted and a record made of the amount, which must be deducted from the total observed weight of the trial.

If it is necessary to reverse the position of the lever or the relative direction of the motion of the pulley (as shown in Figs. 51 and 52), then the weight of the lever must be added to the weight shown by the scale under trial. When the platform scale is used the weight of the lever must necessarily be downward and should be deducted from the weight shown by the

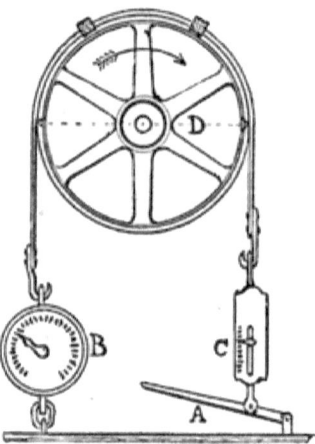

FIG. 53.—DIFFERENTIAL STRAP BRAKE.

scale under trial. Making D equal the diameter of the face of the pulley, fly-wheel, or shaft upon which friction is applied, B the length of the lever from the centre of the shaft to the point of the scale suspension, A the radius of the pulley fly-wheel or shaft, and R the number of revolutions of the shaft per minute: the weight used in the formula must be the net weight of the power stress, or the gross observed weight less the weight of the lever. Then

$$\frac{D \times 3.1416 \times R \times \frac{B}{A} \times \text{weight}}{33,000} = \text{horse-power},$$

or

$$\frac{B \times 6.2832 \times R \times W}{33,000} = \text{horse-power}.$$

$\frac{B}{A} \times \text{weight} = $ the stress or pull at the face of the pulley, and

D × 3.1416 × R = the velocity of the face of the pulley or of the belt that it is to carry.

In Fig. 53 is represented a simple and easily arranged differential strap brake or dynamometer for small motors of less than two horse-power. It consists of a piece of belting held

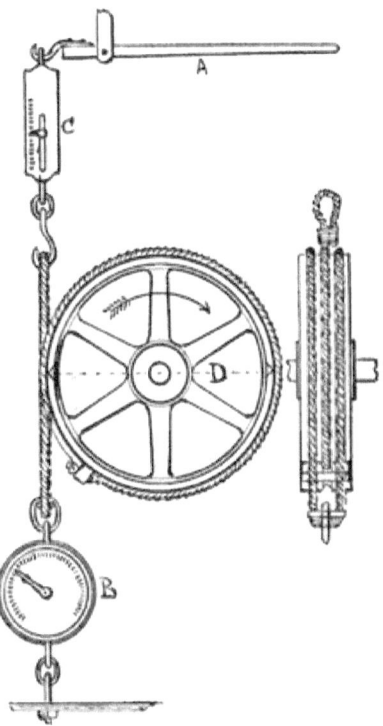

FIG. 54.—DIFFERENTIAL ROPE BRAKE.

in place on the pulley by clips or only strings fastened parallel with the shaft to keep the belt from slipping off; two spring scales, one of which is anchored and the other attached to a hand lever to regulate the compression of the belt upon the surface of the pulley, when the differential weight, B − C, on the scales may be noted simultaneously with the revolutions of the pulley. The simple formula

$$\frac{D \times 3.1416 \times R \times \text{differential weight}}{33,000} = \text{horse-power.}$$

Fig. 54 illustrates a rope absorption dynamometer or brake with a complete wrap on the surface of the pulley, very suitable for grooved pulleys or fly-wheels used for rope transmission. In this form the friction tension may be regulated with a lever as at A. The weight (W) in the formula is the differential of the opposite tensions of the two scales, or B−C=W, Fig. 54, and the formula will then be: $\frac{D \times 3.1416 \times R \times W}{33,000}$ = horse-power, as in the notation, Fig. 53.

Thus it may readily be seen that the difference of the pull in a rope or belt on the two sides of a pulley, multiplied by the velocity of the rim in feet per minute, and the product divided by 33,000, gives the horse-power either absorbed or transmitted by the rope.

The Measurement of Speed.

The revolutions of a motor may be readily obtained by an ordinary hand counter with watch in hand to mark the time; but for accurate work and to show the variations in the flywheel speed by the intervals of revolution between impulses, and especially the effect of mischarges or impulses due to governing the speed, there is no more accurate method than by the use of the centrifugal counter or tachometer.

These instruments are designed to show at a glance a continuous indication of the actual speed and its variation within 2 per cent. by careful handling of the instrument. The tachometer (Fig. 55) with a single dial scale 3 inches in diameter, reading from 100 to 1,000 revolutions per minute, and by changing the gear for the range of gas-engine indication the actual revolutions will be one-half the indicated revolutions, and each divided by 2 will represent the actual speed. In this manner

THE MEASUREMENT OF POWER.

a very delicate reading of the variation in speed may be obtained. For testing the variation of speed in electric-lighting

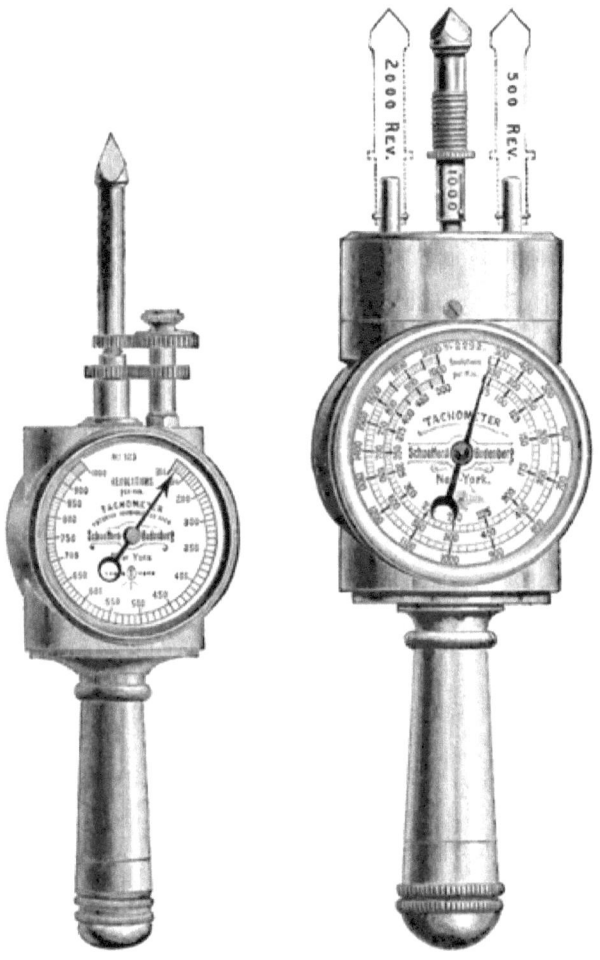

FIG. 55.—THE TACHOMETER.   FIG. 55A.—THE TRIPLE INDEXED TACHOMETRE.

plants operated by gas or gasoline engines, there is no method so satisfactory as by the use of the tachometer.

The triple indexed tachometer (Fig. 55A) is a most convenient instrument for quickly testing and comparing speed of great differences, as the motor and the generator, by simply changing the driving point from one to another gear stem. These tachometers are made by Schaeffer & Budenberg, New York, and may be ordered for any range of speed, from 50 to 500 for gas engines and from 500 to 2,000 for generators, in the same instrument or separate as desired.

The Indicator and Its Work.

We have selected among the many good indicators in the

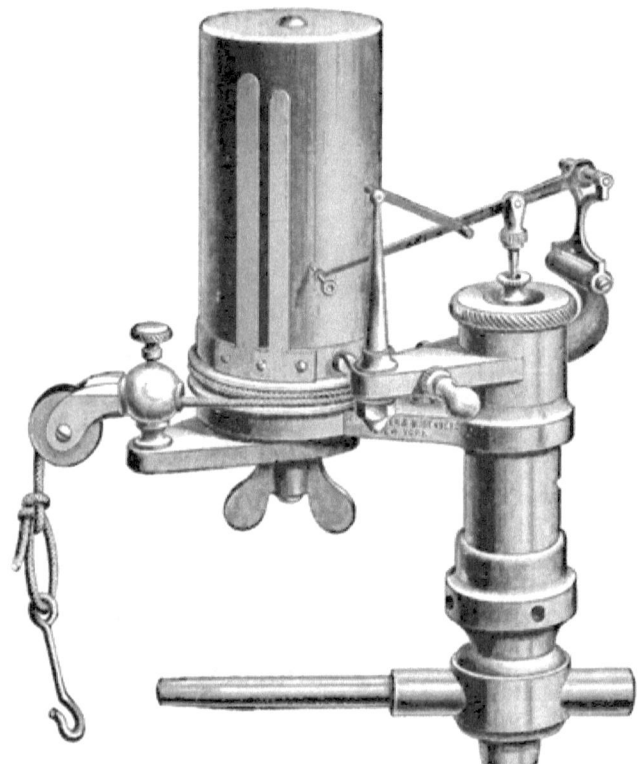

FIG. 56.—THE THOMPSON INDICATOR.

market the one most suitable for indicating the work of the explosive engine. The Thompson indicator as made by

Schaeffer & Budenberg, New York, and illustrated in Figs. 56 and 57, is a light and sensitive instrument with absolute rectilinear motion of the pencil with its cylinder and piston, made of a specially hard alloy which prevents the possibility of sur-

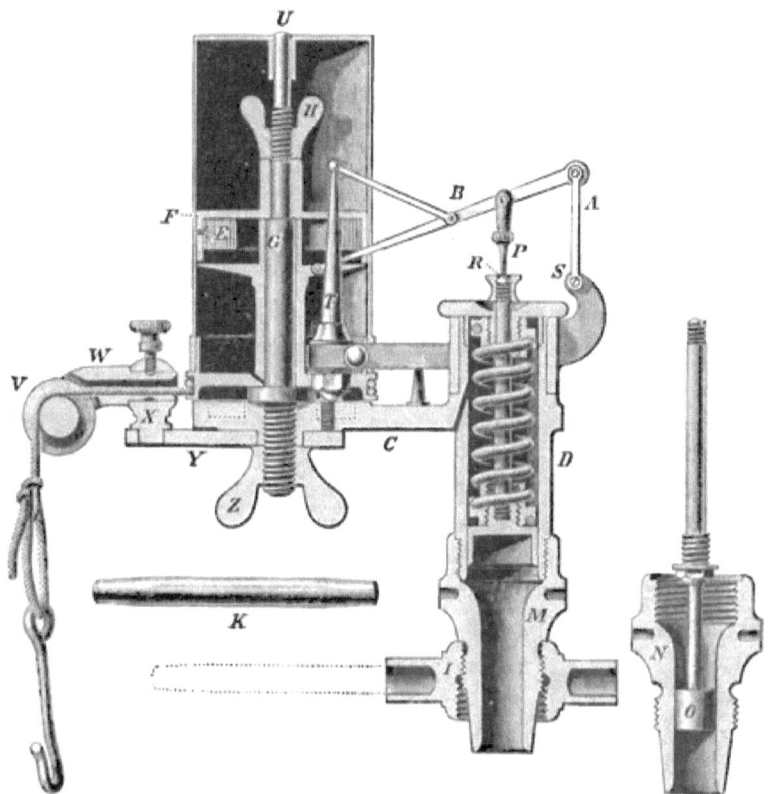

FIG. 57.—SECTION OF INDICATOR.   FIG. 58.—SMALL PISTON.

face abrasion and insures a uniform frictionless motion of the piston. It is provided with an extra and smaller-sized cylinder and piston, suitable with a light spring for testing the suction and exhaust curves of explosive motors, so useful in showing the condition and proportion of valve ports.

The large piston of the standard size is 0.798 inch in diameter and equal to ½ square inch area. The small piston (Fig. 58) is 0.590 inch in diameter and equal to 0.274 square inch area, so that a 50 or 60 spring may be used in indicating explosive engines with the small piston, which will give cards within the range of the paper for low-explosive pressure but full enough to show the variations in all the lines. With the 100 spring and ½ inch area of piston 250 lbs. pressure is about the limit of the card, but with this size piston a 120 or 160 spring is more generally used.

The pulley V is carried by the swivel W and works freely in the post X; it can be locked in any position by the small set screw. The swivel plate Y can be swung in any direction in its plane and held firmly by the thumb-screw Z. Thus with the combination the cord can be directed in all possible directions. The link A is made as short as possible with long double bearings at both ends to give a firm and steady support to the lever B, making it less liable to cause irregularities in the diagram when indicating high-speed motors.

The paper drum is made with a closed top to preserve its accurate cylindrical form, and the top, having a journal bearing at U in the centre, compells a true concentric movement to its surface.

The spring E and the spring case F are secured to the rod G by screwing the case F to a shoulder on G by means of a thumb-screw H.

To adjust the tension of the drum spring, the drum can be easily removed, and, by holding on to the spring case E and loosening screw H, the tension can readily be varied and adapted to any speed, to follow precisely the motion of the engine piston.

The bars of the nut I are made hollow, so as to insert a small short rod K, which is a great convenience in unscrewing the indicator when hot.

The reducing pulley (Fig. 59) is a most important adjunct of the indicator. The revolving parts should be as light as possible and are now made of aluminum for high-speed motors with pulleys proportioned for short-stroke motors. In the use of indicators for high-compression motors it is advisable to have a stop-tube inserted in the cap-piece that holds the spring and extending down and inside the spring so as to stop the motion of the piston at the limit of the pencil motion below the top of

FIG. 59.—THE REDUCING PULLEY.

the card. This will prevent undue stress on the spring and extreme throw of the pencil when by misfires an unusual charge is fired. With the smaller piston and the usual 100 or 120 spring any possible explosive pressure may be properly recorded.

The proximity of the indicator to the combustion chamber is of importance in making a true record of the explosive action of the combustible gases on the card. The time of transmission of the wave of compression and expansion through a tube of one, two, or three feet in length is quite noticeable in the distortion of the diagram. It shows a delay in compression and carries the expansion line over a curve at the apex lower than the maximum pressure, and by the delay raises the expansion curve higher than the actual expansion curve of the cylinder. An indicator for true effect should have a straightway cock screwed into the cylinder.

Vibration of Buildings and Floors by the Running of Explosive Motors.

Since this class of engines has so largely superseded small steam power, and the vast extension of their use in the upper part of buildings due to their economy for all small powers, the trouble arising from vibration of buildings and floors has largely increased.

The necessity for placing motive power near its point of application has resulted in locating gas, gasoline, and oil engines in light and fragile buildings and on floors not capable of resisting the slightest synchronal motion.

This subject has been often brought to our notice since the advent of the gas engine in the lead for small powers. It is a difficult question to advise remedies for it, from the variety of ways in which the effect is produced. Synchronism between the time vibration of a floor and the number of revolutions of the engine is always a matter of experiment, and can only be ascertained by a trial in varying the engine speed by uniform stages until the vibration has become a minimum. Then if the engine speed of least vibration is an inconvenient one for engine economy, or for the speed layout of the machinery plant, a change may be made in the time vibration of the floor by loading or bracing. The placing of a large stone or iron slab under a motor will often modify the intensity of the vibration by so changing the synchronism of the floor and engine as to enable the proper speed to be made with the least vibration.

A vertical post under the engine is of little use unless it extends to a solid foundation on the ground; nor should a vertical post be placed between the engine floor and floor beams above, as it only communicates the vibrations to any floor in unison with the vibrations of the engine floor.

A system of diagonal posts extending from near the centre of a vibrating floor to a point near the walls or supporting columns of the floors above or below, or a pair of iron suspenders placed diagonally from the overhead beams near their wall bearings to a point near the location of an engine and strongly bolted to the floor beams, will greatly modify the vibration and in many cases abate a nuisance.

In the installation of reciprocating machinery on the upper floors of a building in which the reciprocating parts of the motor, as a horizontal engine, are in the same direction as the reciprocating parts of the machines (as in printing pressrooms) the trouble from the horizontal vibration has been often found a serious one. It may be somewhat modified by making the number of the strokes of the engine an odd number of the strokes of the reciprocating parts of the machine.

It is well known to engine builders that explosive motors, like high-speed steam engines, cannot be absolutely balanced, but their heavy fly-wheels and bases go far toward it by absorption, and the best that can be done with the balance is to make as perfect a compromise of the values of the longitudinal and lateral forces as possible by inequality in the fly-wheel rims.

The jar caused by excessive explosions after misfires and muffler-pot explosions is of the unusual kind that cannot be easily provided with a remedy where the transmitted power is not uniform, for where it is uniform there is ample regulation from the governor to make the charges regular, and if the igniter is well adjusted there should be no cause for "kicking," as our European cousins call it. A good practice in setting motors is to locate them near a beam-bearing wall or column that extends to the foundation of the building. Many motors so placed are found to be free from the nuisance of tremor.

CHAPTER XIV.

HEAT EFFICIENCIES.

THE efficiency of an explosive engine is the ratio of heat turned into work in proportion to the total amount of heat produced by combustion in the engine. On general principles the greater difference between the heat of combustion and the heat at exhaust is the relative measure of the heat turned into work, which represents the degree of efficiency without loss during expansion. The mathematical formulas appertaining to the computation of the element of heat and its work in an explosive engine are in a large measure dependent upon assumed values, as the conditions of the heat of combustion are made uncertain by the mixing of the fresh charge with the products of a previous combustion and by absorption, radiation, and leakage. The computation of the temperature from the observed pressure may be made as before explained, but for compression engines the needed starting-points for computation are very uncertain, and can only be approximated from the exact measure and value of the elements of combustion in a cylinder charge.

Then theoretically the absolute efficiency in a perfect heat engine is represented by $\frac{T - T_1}{T}$, in which T is the acquired temperature from absolute zero; T_1, the final absolute temperature after expansion without loss.

Then, for example, supposing the acquired temperature of combustion in a cylinder charge was raised 2000° F. from 60°: the absolute temperature would be $2000 + 60 + 460 = 2520°$, and if expanded to the initial temperature of 60° without loss the absolute temperature of expansion will be $60 + 460 = 520$, then $\frac{2520 - 520}{2520} = .79$ per cent., the theoretical efficiency for the above range of temperature. In adiabatic compression or expansion, the ratio of the specific heat of air or other gases becomes a logarithmic exponent of both compression and expansion. The specific heat of air at constant volume is .1685 and at constant pressure, .2375 for 1 lb. in weight; water = 1. for 1 lb. Then $\frac{.2375}{.1685}$ = the ratio $y = 1.408$.

Then for the following formulas the specific heat $= K_v = .1685$ constant volume, and $K_p = .2375$ constant pressure.

The quantity of heat in thermal units given by an impulse of an explosive engine is, $K_v (T - t)$ = heat units. Then using the figures as before, $.1685 \times (2520 - 520) = 337$ heat units per pound of the initial charge.

The heat in thermal units discharged will be $K_p (T_1 - t)$, $T_1 = t \left(\frac{T}{t}\right)^{\frac{1}{y}}$; t = absolute initial temperature, say $520°$.

Then using again the figures as before and assuming that $T = 2,520° F.$, then $T_1 = 520 \left(\frac{2520}{520}\right)^{\frac{1}{1.408}} = 520 \times$ (log. $4.846 \times .7102$) $= 1594°$ absolute, and $1594 - 520 = 1074° F.$ Then the heat in thermal units discharged will be $.2375 \times (1594 - 520) = .2375 \times 1074 = 255$ heat units.

With the absolute temperature at the moment of exhaust known, the efficiency of the working cycle may be known, always excepting the losses by convection through the walls of the cylinder.

The formula for this efficiency is: eff. $= 1 - y \frac{T_1 - t}{T - t}$; then by substituting the figures as before, $1 - 1.408 \frac{1594 - 520}{2520 - 520} = \frac{1074}{2000} = .537 \times 1.408 = .756$, and $1 - .756 = 24$ per cent.

To obtain the adiabatic terminal temperature from the relative volumes of clearance and expansion, we have the formula $\frac{V_0}{V}^{-y-1} = \frac{T_1}{T}$, in which $\frac{V_0}{V}$ is the ratio of expansion in terms of the charging space in engines of the Lenoir type to the whole volume of the cylinder including the charging space, so that if the stroke of the piston is equal to the area of the charging or combustion space, the expansion will be twice the volume of the charging space and $\frac{V_1}{V} = \frac{1}{2}$. Then $\frac{T_1}{T} = \left(\frac{1}{2}\right)^{.408}$ and $T_1 = T\left(\frac{1}{2}\right)^{.408}$. Using the same value as before, $T_1 = 2520 \left(\frac{1}{2}\right)^{.408}$ and using logarithms for $\frac{1}{2}$, log. $2 = 0.30103 \times ^{.408} = $ log. 0.12282 = index 1.32, and $\frac{2520°}{1.32} = 1908°$, the absolute temperature T_1 at the terminal of the stroke. Then $1908° - 460° = 1448°$ F., temperature at end of stroke.

For obtaining the efficiency from the volume of expansion from a known acquired temperature we have $\frac{V}{V_0} t = \frac{2}{1} \times 520°$ = $1040°$ absolute = t_1. Then

$$\text{the efficiency} = \frac{1. - (T_1 - t_1) + y(t_1 - t)}{T - t}.$$

Then using the values as above,

$$\text{efficiency} = \frac{1. - (1908 - 1040) + 1.408 (1040 - 520)}{2520 - 520} = 868 +$$

$1.408 \times 520 = 732 + 868 = \frac{1600}{2000} = .80$, and $1 - .80 = .20$ per cent.

For a four-cycle compression engine with compression say to 45 lbs, the efficiency is dependent upon the temperature of compression, the relative volume of combustion chamber and piston stroke, and the temperatures. Fig. 60 is a type card of reference for the formulas for efficiencies of this class of explosive motors, in which:

        t = abs. temp. at b normal.
        t_0 = abs. temp. of compression f.
        T = abs. acquired temp. e.
        T_1 = abs. temp. at c.
        P = abs. pressure at b.
        P_e = abs. pressure at f.
        P_0 = abs. pressure at c.

HEAT EFFICIENCIES.

V_e = volume at b.
V = volume at c.
V_a = volume at f.
vo = V or volume at compression = volume at exhaust.
K_v = .1685 specific heat at constant volume.

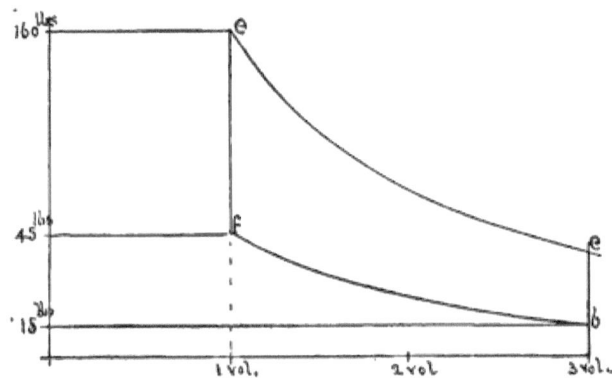

FIG. 60.—THE FOUR-CYCLE COMPRESSION CARD.

Let T = abs. acquired temp. = 2520° F. as before.
t = abs. normal temp. = 520° or 60° F.

$$t_e = \text{abs. temp. of compression} = t\left(\frac{P_e}{P}\right)^{\frac{\gamma-1}{\gamma}} = \frac{1.408-1}{1.408}$$

$$= 0.29. \text{ Then } 520°\left(\frac{60}{15}\right)^{0.29} = 777° \text{ absolute.}$$

T_1 = abs. temp. of expansion = $\frac{T\,t}{t_e}$ or $\frac{2520° \times 520}{777}$ = 1686°.

The terms being assumed and known from assumed data, the efficiency = $1 - \frac{K_v(T-t_e) - K_v(T_1-t)}{K_v(T-t_e)}$.

Reducing, efficiency = $1 - \frac{T_1-t}{T-t_e}$; substituting figures as above found, $1 - \frac{1686-520}{2520-777} = .333$ per cent.; also $1 - \frac{T_1}{T} = \frac{1686}{2520} = .333$ and $1 - \frac{t}{t_e} = \frac{520}{777} = .333$.

For obtaining the efficiency from the relative volumes at both ends of the piston stroke, with an expansion in the cylinder equal to twice the clearance space, by which the total volume at the end of the stroke will be three times the volume of the clearance space,—the efficiency in this case may be expressed by the formula $1 - \left(\dfrac{V_0}{V_0}\right)^{y-1}$; substituting, the values become $1 - \left(\dfrac{1}{3}\right)^{.408}$; using logarithms as before, log. $3 = 0.477121 \times .408 = 0.194665$, the index of which is 1 565, and $\dfrac{1}{1.565} = .639$. Then $1 - .639 = .36$ per cent.

CHAPTER XV.

EXPLOSIVE ENGINE TESTING.

For the reason that elaborate and complicated tests have been made and exploited in other works on the gas engine, which may be referred to for the details of expert work, the author of this work has decided to reduce the practice of testing explosive motors to a commercial basis on which purchasers can comprehend their value as a business investment for power. The disposition of builders of explosive engines to follow the economics in construction in regard to least wall surface in contact with the heat of combustion, and of maintaining the wall surface at the highest practical temperature for economical running by the rapid circulation of warm water from a tank or cooling coil, leaves but little to accomplish, save the proper size and adjustment of the valves and igniters for the engines, in order that they may properly perform their functions. The indicator card, if made through a series of varying proportions of gas or gasoline and air mixtures, will show the condition of the adjustments for economic working. The difference between the indicated power for the gas used by the card and the power delivered to the dynamometer or brake shows the mechanical efficiency of the engine. The best working card of the engine should be a satisfactory test to a purchaser that the principles of construction are correct. A brake-trial certificate or observation should satisfy as to frictional economy, and the price and quantity of gas per horse-power hour should settle the comparative cost for running. The variation in the heating power of illuminating gas in the various parts of the United States is much less than its variation in price. Producer gas is a specialty for local consumption, and its cost drops with its heating power.

Apart from the actual cost of gas in any locality and the quantity required per brake horse-power, durability of a motor is one of the principal items in the purchase of power.

In the use of gasoline, kerosene, and crude petroleum in explosive engines, their heating values are uniform for each kind, and as motors are generally adjusted for the use of one of the above hydrocarbons only, the difference of cost between these various fuels is the best indication as to the relative cost of power.

No instruments have yet been contrived for giving the temperatures of combustion, either initial or exhaust, in an internal combustion motor; for at the proper working speed the changes of temperature are so rapid that no reliable observation can be made even with the electric thermostat, as has been tried in Europe. The computed temperatures are unreliable and at best only approximate; hence the indicator card becomes the only reliable source of information as to the action of combustion and expansion in the cylinder, as well as to the adjustment of the valves and their proper action.

The temperature of combustion as indicated by the fuel constituents, and computed from their known heat values, gives at best but misleading results as indicating the real temperature of combustion in an explosive engine. There is no doubt that the computed temperatures could be obtained if the contaminating influence of the neutral elements that are mixed with the fuel of combustion, as well as the large proportion of the inert gases of previous explosions, could be excluded from the cylinder, when the radiation and absorption of heat by the cylinder would be the only retarding influences in the development of heat due to the union of the pure elements of combustion.

For obtaining the indicated horse-power of a gas, gasoline, or oil engine, the mean effective pressure as shown by the card may be obtained by dividing the length of the card into ten or any convenient number of parts vertically, as shown in Fig. 61 for a four-cycle compression engine. For each section measure the average between the curve of compression and the curve of expansion with a scale corresponding with the number of the indicator spring. Add the measured distances and divide

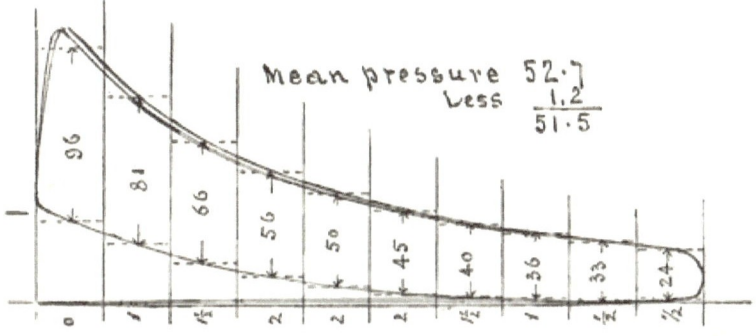

FIG. 61.—FOUR-CYCLE GAS-ENGINE CARD.

by the number of spaces for the mean pressure. With the mean pressure multiply the area of the cylinder for the gross pressure. If there have been no misfires, then one-half the number of revolutions multiplied by the stroke and by the gross pressure, and the product divided by 33,000 will give the indicated horse-power. If there is any discrepancy along the atmospheric line by obstruction in the exhaust or suction stroke, the average must be deducted from the mean pressure.

The exhaust valve, if too small or with insufficient lift, or a too small or too long exhaust pipe, will produce back pressure on the return line, which should be deducted from the mean pressure. A small inlet valve or too small lift, or any obstruction to a free entry of the charge, produces a back pressure on the outward or suction stroke and a depression along the atmospheric line, which must also be deducted from the mean pressure.

It is assumed that the taking of an indicator card must be done when the engine is running steady and at full load. During the moment that the pencil is on the card there should be no misfires recorded, in order that the card may represent the true indicated horse-power of the engine. The record of the speed of the engine should be taken at the same time as the card, but the measurement of the quantity of gas used cannot be accurately observed on the dial of an ordinary gas meter during the few moments' interval of the card record and speed count. For the gas record, the engines should be run at least five minutes at the same speed and load and an exact count of the explosions made. The misfires or rather mischarges in an engine running with a constant load are of no importance in the computation for power because they are properly caused by overspeed, and the overspeed and underspeed should make a fair balance for the average of the run as indicated by the speed counter.

The number of cubic feet of gas indicated by the meter for a few minutes' run, multiplied by its hour exponent and divided by the indicated power by the card or the actual horse-power by the brake, will give the required commercial rating of the engine as to its economic power. The difference as between the cost of gas for the igniter and the cost of electric ignition is too small to be worthy of consideration.

In testing with gasoline or oil the detail of operation is the same as for gas, with the only difference of an exact measure of the fluid actually consumed in an hour's run of the engine under a full load. The loading of an engine for the purpose of testing to its full power is not always an easy matter; although, when driving a large amount of shafting and steady-running machines, a brake may be conveniently applied to increase the work of the engine. In trials with a brake alone, a continual run involves some difficulties on account of the intense friction and heat produced, which makes the brake power vary considerably and cause a like variation in the ignitions.

This only becomes serious when temporary brakes have to be improvised, but in engine-building establishments brakes are used that are specially designed for uniform resistance and continued testing.

CHAPTER XVI.

VARIOUS TYPES OF ENGINES AND MOTORS.

The Economic Gas Engine.

MANY of the engines of the Economic Gas Engine Company are still in use. We illustrate their design as being one of the earlier types in use in the United States. It is of the two-cycle

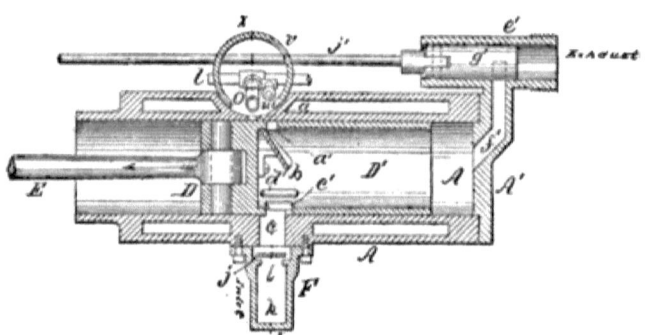

FIG. 62.—SECTION OF CYLINDER.

non-compression type of Lenoir, with an indicator card of the form shown in Fig. 3. A section of this engine is shown in Fig. 62, in which A is the jacketed cylinder, D the piston with an elongated shell D', F air and gas inlet and mixer, J a check valve; c and c' mixed gas and air ports, d auxiliary air port; g' piston exhaust valve with exhaust port f', b a deflector and a' firing port.

The operation is as follows: The piston sweeps the products of a previous combustion out at the exhaust port by the piston following to a point when the inlet ports in the piston are just past the inlet ports in the cylinder, when the exhaust port closes and the suction of a charge commences and is continued

FIG. 63.—THE ECONOMIC PUMPING ENGINE.

FIG. 64.—THE ECONOMIC

until the inlet ports are closed by the outward stroke of the piston. At this point the firing ports of cylinder and piston are in line and the explosion takes place with all the ports closed to the end of the impulse stroke, when the exhaust port opens by a cam and the products of combustion are again swept out

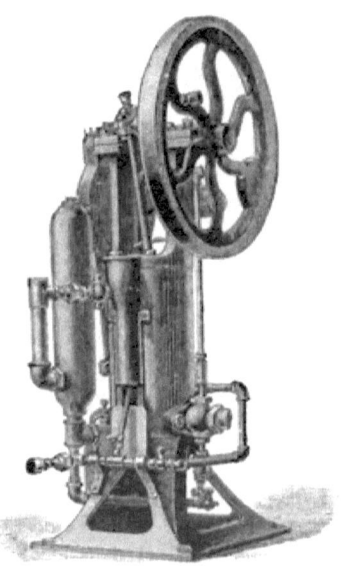

FIG. 65.—THE VERTICAL PUMPING ENGINE.

with the exception of the clearance space within the shell of the piston.

This engine, like others of its type made in Europe, is not considered economical as compared with the later engines of the four-cycle compression type. The various designs as made by different makers consume from 80 to 50 cubic feet of illuminating gas per horse-power hour, the latter figure being the rate for the Economic as made ten years since.

The New Era Gas Engine

is of the four-cycle compression type with a heavy and substantial base. The valve-gear shaft being driven by a worm

VARIOUS TYPES OF ENGINES AND MOTORS. 115

gear from the main shaft, insures a smooth and noiseless motion. The illustration (Fig. 66) on this page has one of the fly-

FIG. 66.—THE NEW ERA GAS AND GASOLINE ENGINE.

wheels left off to show the arrangement of the worm gear, which is also shown in Fig. 67 in detail. This method of driving the valve-gear shaft is fast growing in favor, and is now largely in use.

The valves are of the poppet type, operated by cams on the secondary shaft, which also drives the governor through bevel-

FIG. 67.—THE WORM GEAR.   FIG. 68.—VALVE CHEST.

speed gear. All the valve chambers have flanged plugs for facilitating the removal and cleansing of the valves.

The end view of the lateral shaft and valve chest with the

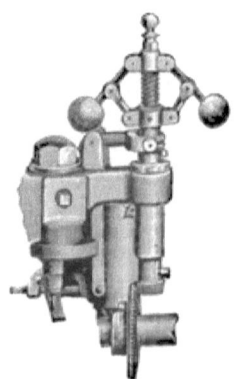

FIG. 69.—THE GOVERNOR.

attachment of the tube igniter is shown in Fig. 68. The electric igniter is applied at the same opening in the valve chest as used for the tube igniter.

The governor is of the ball type, running direct from the secondary shaft by a bevel gear, and through a bell-crank lever and arm controls the gas-inlet valve. Fig. 69 shows the arrangement more in detail and also the great convenience in gas engines, a cap plug for quickly removing the valve and an inspection plug at the side of the valve chest.

The fuel for these engines may be illuminating gas, producer gas, natural gas, or gasoline. The cost for running can be gauged only by the quantity, say 15 to 20 cubic feet illuminating gas or one-tenth of a gallon of gasoline per indicated horse-power per hour.

In using gasoline a small pump (Fig. 70) is attached to the

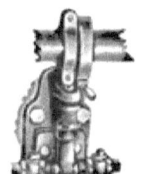

FIG. 70.—THE PUMP.

engine bed and driven by a cam on the lateral shaft. The pump draws from a tank set in a safe place, underground if possible and draws a few drops of gasoline at a stroke, forcing it into the air chamber, where it is vaporized and mixed with the incoming air. The surplus, if any, is returned to the tank. These engines are made in sizes from 10 to 50 B.H.P.

The Pierce Gas and Gasoline Engine.

This engine is built on the four-cycle compression type, as shown in the illustrations of both sides of the 1 to 5 H.P. engines (Figs. 71 and 72). This company also build engines of 6, 8, 10, 12, 15 and 20 H.P. These figures represent the brake or actual horse-power.

The valve motion is taken from the main shaft with spur gears and secondary shaft upon which there is a cam that operates the valves through a connecting rod. On the face of the cam is a wrist pin, carrying a connecting rod, which operates both the governor and the electrical firing device.

The poppet valves never require oil; they lift squarely from

FIG. 71.—THE PIERCE GAS AND GASOLINE ENGINE—RIGHT SIDE.

their seats. They wear smooth and bright and are easily uncovered for regrinding when necessary. The entire operating mechanism is in plain sight and all wearing parts can be readily examined and adjusted without removing or taking the engine apart. The governor is very simple and sensitive. It is composed of three pieces: a hardened steel finger, weighted and held to its proper position by an adjustable spring. The weighted finger acts as an inverted pendulum swung by the movement of the connecting rod, making a miss gas charge when the engine speed is too high. It is adjustable by mov-

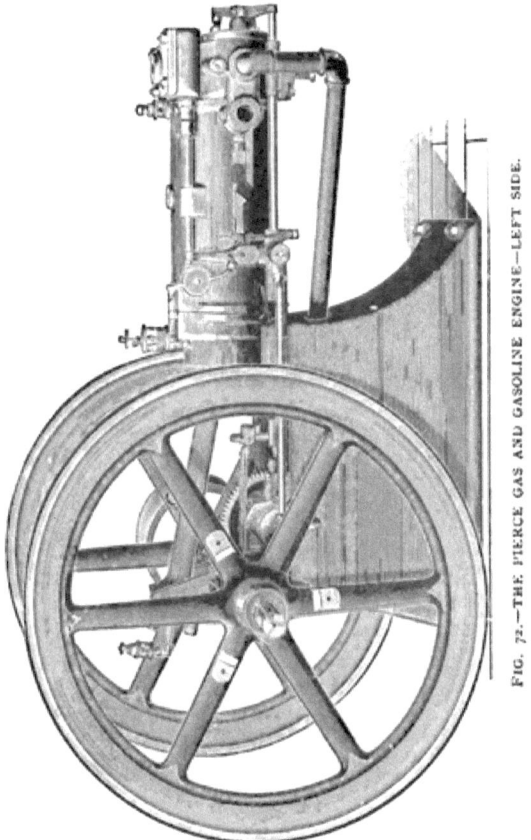

FIG. 72.—THE PIERCE GAS AND GASOLINE ENGINE—LEFT SIDE.

ing the weight on the stem and by a spiral spring and adjusting nut. These engines are built to run with coal gas, natural gas, and gasoline, can be changed from one fuel to another with little trouble, and are also made to change while the engine is running.

The electrical firing device is very simple. It is composed

of two electrodes, one a flat piece of steel $\frac{1}{4}$ inch wide by $\frac{3}{8}$ inch long and $\frac{1}{16}$ inch thick. The other is a piece of No. 16 wire. One is insulated from the engine and the other in circuit with it. A make-and-break spring at the side of engine (also insulated from the frame) forms the circuit when the electrodes come together. In parting the spark is made which fires the charge. The electrodes never corrode, as they clean themselves every time they pass each other, and they will remain clean until they are worn out. A four-cell battery is used and will run these engines 1,800 hours without recharging.

Cost of Operation.—These engines run with a consumption of illuminating gas of 16 cubic feet per actual horse-power per hour; with gasoline, $\frac{1}{10}$ of a gallon per actual horse-power per hour.

For the use of gasoline, a small pump is attached to the engine, which pumps the gasoline to a small cup from a tank placed underground or in a safe place; from the cup the gasoline is fed directly to the cylinder air inlet. If more gasoline is pumped than required, the excess runs back to the tank; 0.74 gravity gasoline is used.

The Charter Gas and Gasoline Engine.

The Charter is a representative of one of the earliest types of American gas engines. It has gone through its evolution of improvement, and claims to be a model of simplicity. It is of the four-cycle compression type. It runs equally well with illuminating gas, natural gas, and gasoline. It is built in nine sizes, from $1\frac{1}{4}$ to 35 B.H.P. The cut (Fig. 73) represents five sizes, and Fig. 74 represents the smallest size, No. 00, which is vertical and of $1\frac{1}{4}$ B.H.P. Both tube and electric ignition are used with these engines. In the horizontal engine the mixing chamber is attached to the head of the cylinder, into which the gas or gasoline is injected by the operation of the small pump G (Fig. 75), driven by a rod and levers operated by a cam on the secondary shaft. The nozzle H (Fig. 75) projects upward so that the indraught from the air pipe N supplies the required quantity, while the overplus is returned to the tank when placed below the engine. When the gasoline tank is placed above the engine so that there is a gravity flow to the engine, the flow is regulated by two valves

FIG. 73.—THE CHARTER GAS AND GASOLINE ENGINE.

in the flow pipe, a throttle valve at the pump, and by the operation of the plunger of the pump, which in this case does not force a specific quantity of gasoline, but only opens the way for an instant of time to a flow produced by gravity and the suction of the cylinder. In this arrangement, any stoppage of the engine other than by closing the gasoline valves will stop the flow of gasoline by the covering of the pump ports by the plunger. The governor is of the centrifugal type, mounted on the pulley, and consists of two balls held in tension by springs, which operate a sleeve on the main shaft through a bell-crank movement. The movement of the sleeve throws the injector-rod roller on to or off the cam on the secondary shaft, thus making a "hit or miss" injection from the pump.

Communication between the mixing chamber and the cyl-

FIG. 74.--THE VERTICAL CHARTER.

inder is cut off, at the moment the charge to the cylinder is completed and compression commenced, by a gravity-poppet valve at B (Fig. 75). The operation of the pump plunger is the same for gas as for gasoline: the plunger only opening a way for the flow of the gas at the proper moment, and being governed in its operation the same as when gasoline is used. The exhaust valve is of the poppet type, operated by a cam on the secondary shaft, the movement of which also operates the oil cup on the cylinder by the levers and small-rock shaft, as shown in Fig. 75. The detail of the operating parts are well

VARIOUS TYPES OF ENGINES AND MOTORS.   123

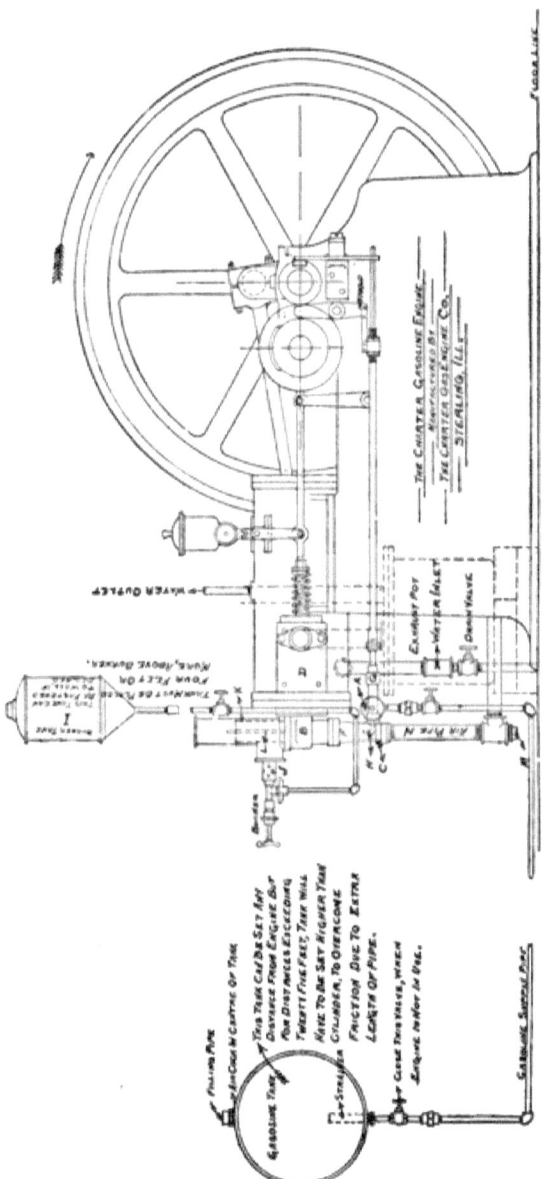

FIG. 75.—THE CHARTER—DETAILS OF GASOLINE CONNECTIONS.

shown in the skeleton cuts of the horizontal and vertical engines (Fig. 76 and Fig. 77). A relief valve for easy starting is placed on the cylinder of No. 2 and larger engines. The No. 6 and No. 7 engines are furnished with a perfect and practical starter. The ignition-tube burner is shown in the different illustrations, consisting of a gas or gasoline jet in a

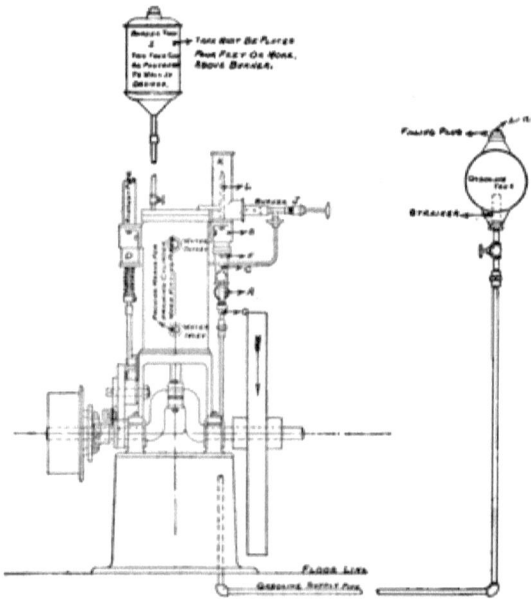

FIG. 76.—THE VERTICAL CHARTER FOR GASOLINE.

perforated sleeve, acting as a Bunsen burner upon the compression tube contained in the asbestos-lined chimney.

For electric ignition a pair of insulated electrodes in a plug are screwed into the place of the tube igniter and operated by a spark breaker.

The Charter Gasoline Pumping Engine.

Fig. 79 shows an engraving of the Charter gasoline engine and pump combined. This combination was designed for any kind of service that piston pumps are capable of. It is compactly built, a feature which, in places where floor space is valuable, is especially desirable. It is easily operated. When

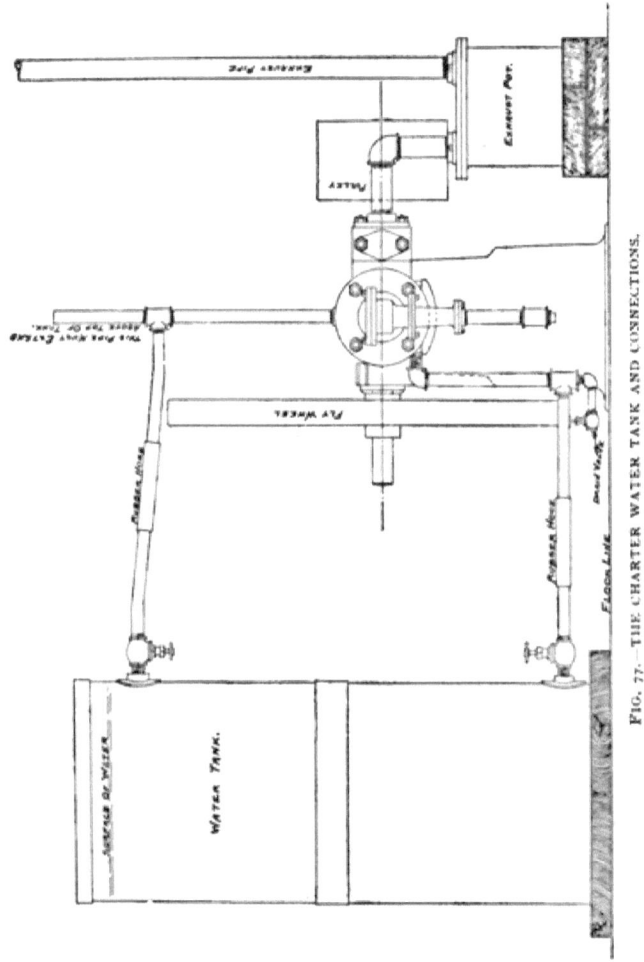

FIG. 77.—THE CHARTER WATER TANK AND CONNECTIONS.

through pumping, nothing remains to do but shut off the gasoline. As no special attendant is required, it is especially desirable for filling railroad tanks, as the station agent or his assistant can take care of the engine and see that the pumping is done without interfering with their regular duties, thus saving the expense of employing a man to go from station to station

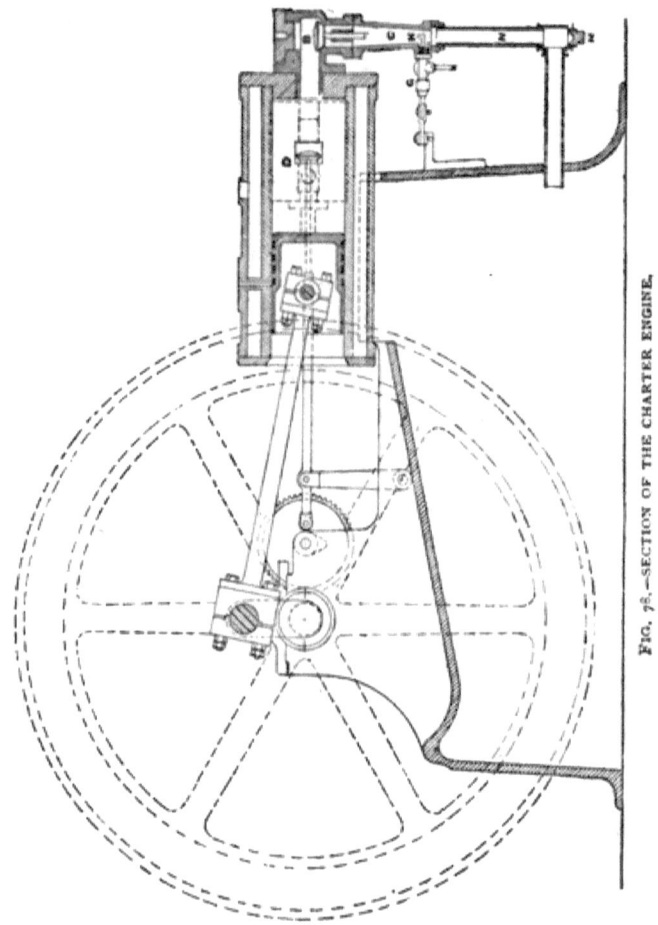

FIG. 78.—SECTION OF THE CHARTER ENGINE.

to fill the tanks. It is a suitable pumping engine for hydraulic elevators. The gears are all machine cut, the pump cylinder is brass lined, and everything about the engine and pump is built on the interchangeable plan. The cut illustrates an engine and pump capable of delivering 60 gallons of water per minute against 100 or 200 feet head, or equivalent pressure. It is self-contained and may be set in operation almost any-

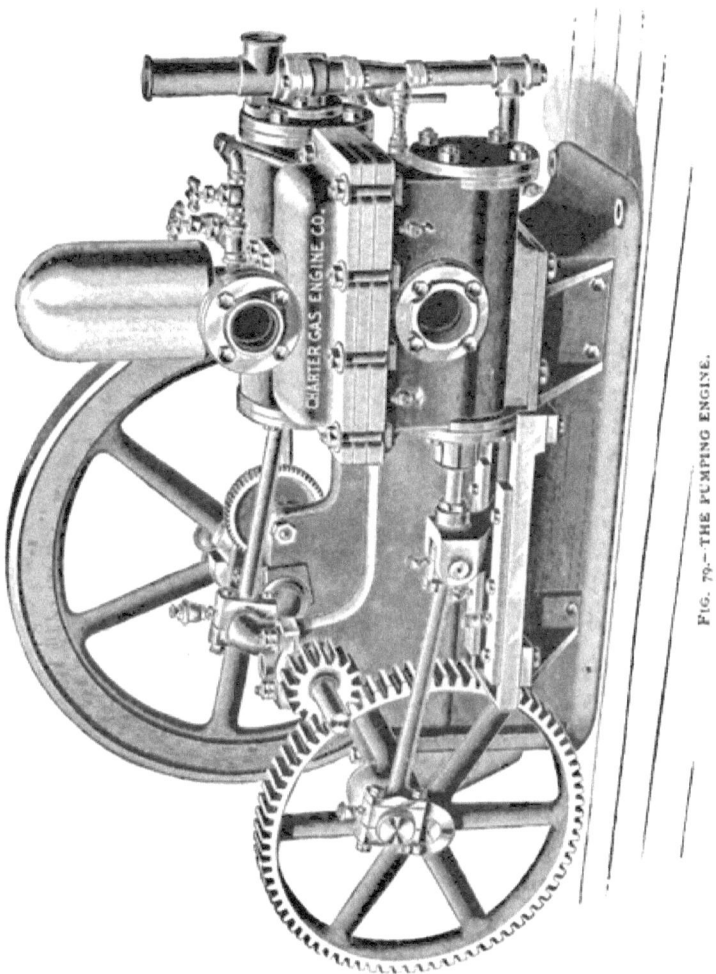

FIG. 79.—THE PUMPING ENGINE.

where. The pump gear is easily detached and a pulley supplied for temporary power use, making this combination a valuable one for agricultural work and irrigation.

The Raymond Gas and Gasoline Engines.

These engines are built in three styles, all in the vertical four-cycle compression type. The quadruple engine (Fig. 80), in

FIG. 80.—THE RAYMOND QUADRUPLE GAS ENGINE.

which there are two impulses during each revolution of the shaft, are made in three sizes: 60, 85, and 100 H.P. (actual).

The duplex (Fig. 81) with a section view (Fig. 82), in which one impulse is made for each revolution, are made in ten sizes, from 4 to 50 H.P. (actual).

The details of construction are similar in all the styles and

FIG. 81.—THE DUPLEX RAYMOND.

sizes. They are entirely enclosed in a base with a vent pipe at the back to prevent cushioning by the pistons, and, with the large flange on the front of the base, are removable for easy feed-oil access to the moving parts within.

The valves are of the rotating type and are operated directly from the crank shaft by a set of bevel and spur gear;

they are held to their seats by spiral springs and are supplied with steel ball bearings. The valves are lubricated from sight feed oil-cups.

Fig. 82 shows a section of one of the cylinders of a duplex

FIG 82. SECTION OF THE DUPLEX RAYMOND.

with the bevel gear, secondary shaft, and spur wheels of the valve gear.

The governor is placed on the fly-wheels, and is of the centrifugal type, and regulates through piston valves the exact amount of gas or gasoline mixture required for each impulse to maintain a perfectly steady speed of engine under all conditions and variations of load.

For the use of gasoline, naphtha, or light petroleum oil, a glass reservoir is placed on top of the vaporizer of the capacity of a half-pint which is connected to a small pump, which in turn is connected to a gasoline tank.

A return pipe connects the reservoir with the tank for return of the surplus gasoline. The adjustable needle valve,

FIG. 83.—THE RAYMOND, SINGLE CYLINDER.

which governs the supply of gasoline necessary to give the engine its required power and steady motion, is in direct connection with the shaft governor and works automatically.

The hot and cold air valve, or air mixer, connects the vaporizer with a jacket around the exhaust pipe, in which the air is heated to more effectually vaporize the gasoline. An explosive starter is provided for the large engines.

Fig. 83 illustrates the Raymond single cylinder engine for

gas, gasoline, or light oil, showing the cover removed to expose the valve gear and adjustable spring for tightening the rotating valve. It is made in ten sizes, from 1 H.P. (actual) to 20 H.P. (actual).

It is claimed that an economy of 12 cubic feet of natural gas per actual horse-power has been attained, and a guaranty of 15 cubic feet per actual horse-power is made.

The Sintz Gas Engine.

This engine is of the two-cycle compression type, taking an impulse at every revolution, yet it is different from the usual

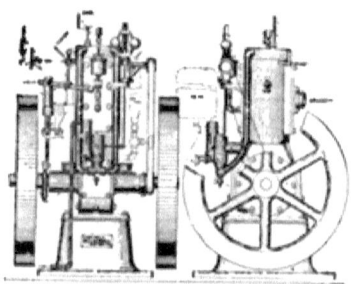

FIG. 84.—THE SINTZ ENGINE.

action of the ordinary two-cycle non-compression type, for it is a compression engine with enclosed crank and piston connections, so that with the up-stroke of the piston air is drawn into the crank casing and by the return stroke the air is slightly compressed. When the down-stroke of the piston nears the terminal, it opens an exhaust port in one side of the cylinder, and at a little farther advance of the piston opens an inlet port on the other side of the cylinder, through which the compressed air in the crank chamber rushes to charge the cylinder, at the same time the gas valve is opened by the eccentric; or if gasoline is used, the pump injects a charge of gasoline in a fine spray at the proper moment. By means of a deflector on the inlet side of the piston, the incoming charge is thrown upward toward the top of the cylinder, thus separating the discharging products of the previous explosion from the fresh charge and by this means obtaining a purer mixture for the next explosion.

The ascension of the piston gives a full compression and time for the mixture to become uniform for ignition by tube or electric igniter. It may be called a valveless engine, as the piston itself opens both the exhaust and inlet ports. A light check valve only is used to check the return of the air drawn into the crank chamber by the upward movement of the piston.

In Fig. 84 is represented the stationary Sintz engine, front and side view. The governor is of the centrifugal type, lo-

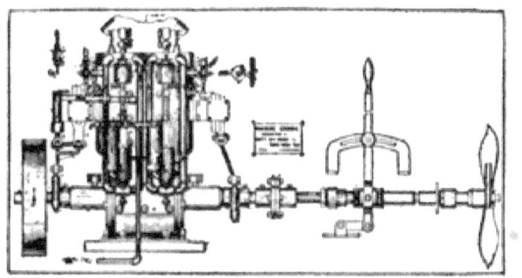

FIG. 85.—THE SINTZ DUPLEX MARINE ENGINE.

cated in the fly-wheel, where two balls held by springs operate through bell-cranks the movement of a sleeve on the main shaft carrying a cam, which by the position of the sleeve determines the operation of the cam on the gas valve, or on the gasoline pump when gasoline is used. The cam is so constructed as to regulate the flow of gas or gasoline to modify the explosive mixture, and not by the entire suspension of an explosion.

Fig. 85 shows the duplex marine engine with its reversing propeller. The reversing gear operated by the lever contains all the movements required for full head, slowing, dead centre, slow backing, and full back—one of the neatest arrangements yet made for the management of boats driven by gas engines. Other arrangements of the reversing lever are made so as to place it in the forward part of the boat with the steering gear.

A section of the Sintz cylinder (Fig. 86) shows somewhat in detail the inlet and exhaust ports with the deflector on the piston opposite the inlet port. The compressed air port in a recess in the lower part of the cylinder shuts off a portion of

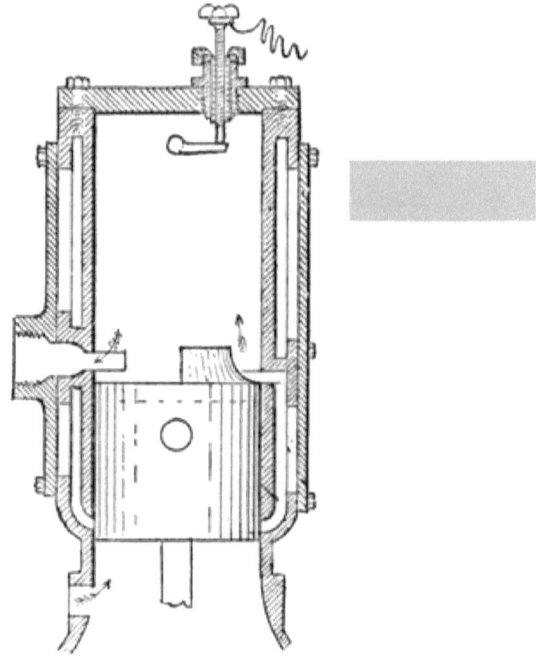

FIG. 86.—THE CYLINDER.

the compressed air at the moment that the inlet port opens, by which means a measured charge of fresh air is forced into the cylinder at every revolution of the shaft. The slight compression by the down-stroke of the piston is sufficient to charge the air chamber in the cylinder for an explosion charge by its expansion through the inlet port during the part of the crank revolution due to the amount of port opening.

The electrode entering at the top through the cylinder cover makes contact and spark break by the rocking arm on a spindle passing through the side of the cylinder. The time regulation is adjusted by the insulated screw electrode, while the break arm is operated by a connecting rod from the pump arm; both pump and breaker are operated by one cam.

In the gasoline stationary engines the required quantity of gasoline is regulated by a needle valve operated by the governor, while in the marine engines the needle valve is operated by a rod extending to the steering wheel. With the extension of the reversing-gear connection to the steering wheel forward, all the operations for running a boat are managed by one person.

The Atkinson Gas Engine.

This unique motor, first brought out in England, and made in the United States by the Warden Manufacturing Company,

FIG. 87.—THE ATKINSON GAS ENGINE.

is of the two-cycle type, in which compression, expansion by combustion, exhaust, and recharging are accomplished by the motion of the piston during each revolution are produced by a toggle-joint movement across the centre line of the engine.

Its cyclical recurrence is seemingly a near approach to an

ideal motor from the fact that the clearance is small in proportion to the volume in the fresh charge, and therefore the explosive effect is much greater than in motors of the four-cycle type.

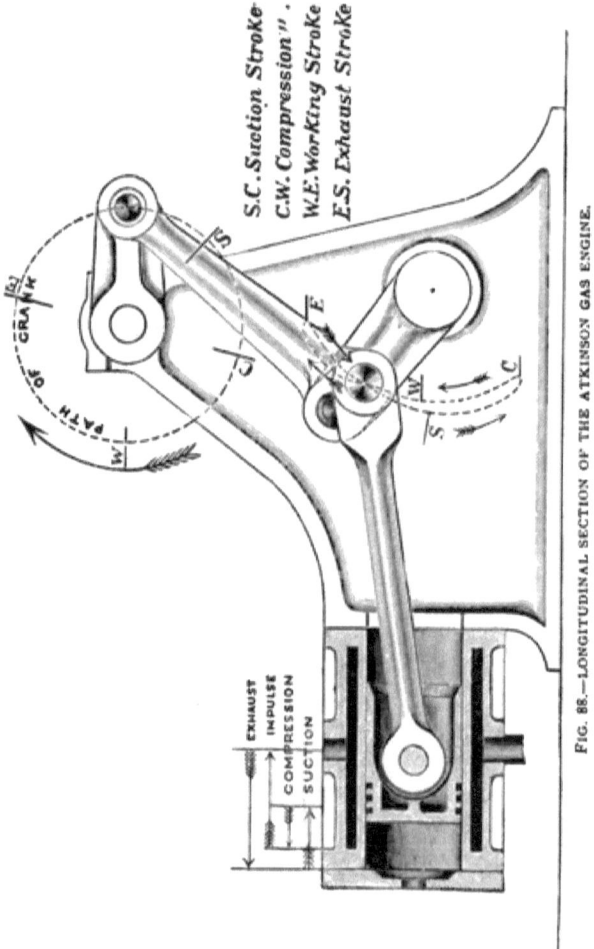

FIG. 88.—LONGITUDINAL SECTION OF THE ATKINSON GAS ENGINE.

Fig. 87 shows a perspective view of the engine, and Fig. 88 is a sectional elevation showing the movement of the toggle connection in producing the four distinct movements of the piston for each revolution of the shaft.

VARIOUS TYPES OF ENGINES AND MOTORS.   137

It will be noticed by a careful inspection of the sectional elevation that the different operations are obtained by the addition of but two parts, a link which vibrates through the arc of

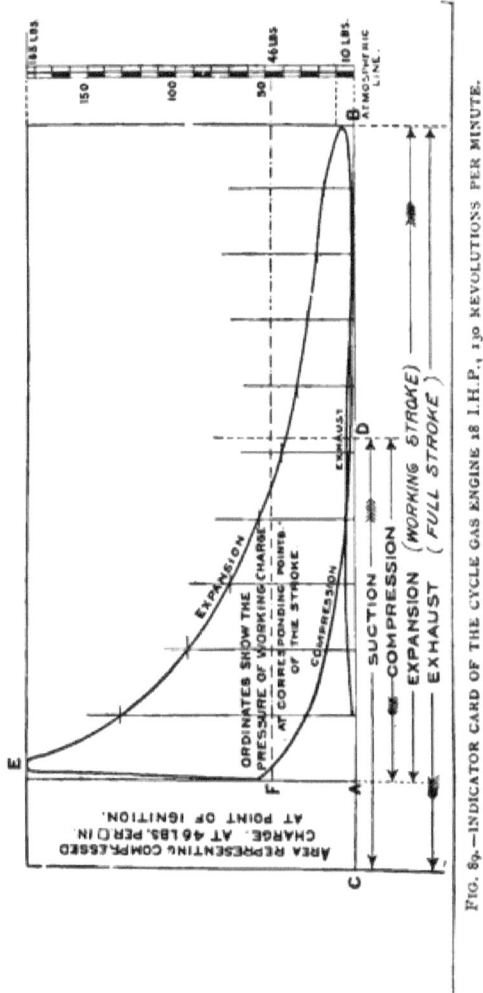

FIG. 84.—INDICATOR CARD OF THE CYCLE GAS ENGINE 18 I.H.P., 130 REVOLUTIONS PER MINUTE.

a circle, a connecting rod, and by changing the position of the crank shaft in relation to the cylinder.

The outer end of the piston connecting rod is attached to a

pin passing through the crank connecting rod, and the latter is connected to the link. The different centres are so placed in relation to each other and to the centre line of the cylinder that the centre of the pin to which the piston connecting rod is attached travels in a curve resembling the figure eight, passing over the portion SC (Fig. 88) during the suction stroke, over CW during the compression stroke, over WE during the working or explosive stroke, and over ES during the exhaust stroke.

The figure shows that the compression stroke is shorter than the suction stroke, that the working stroke is almost double the suction stroke, that the exhaust stroke ends with the piston as close to the cylinder cover as it is possible mechanically to have it, and that the working stroke takes place in one-quarter of a revolution.

The clearance space beyond the terminal exhaust position of the piston is so small that practically the products of combustion are entirely swept out of the cylinder during the exhaust stroke, so that each incoming charge has the full explosive strength due to the mixture used.

It is also possible to expand the exploded charge to such a volume that the terminal pressure will be reduced to the lowest practical point, and that, owing to the purity of the charge, the greatest possible pressure will be attained at the commencement of the expansion.

In Fig. 89 is represented an indicator card taken from an 18 H.P. engine. It is a most interesting study and shows the value of a pure mixture in the quick and sharp terminal of the explosive effect, occupying only about 0.09 of a second in duration and a pressure of 185 lbs. per square inch, with the expansion line falling in good form to 10 lbs. at the exhaust end —the mean pressure being 49 lbs., which is equal to about 80 lbs. mean pressure in a four-cycle engine, considering the difference in idle piston travel and comparative proportion of expansion stroke.

The Webster Gas and Gasoline Engine.

These engines as now made are improvements on the Lewis engine as formerly made. Fig. 90 represents the ver-

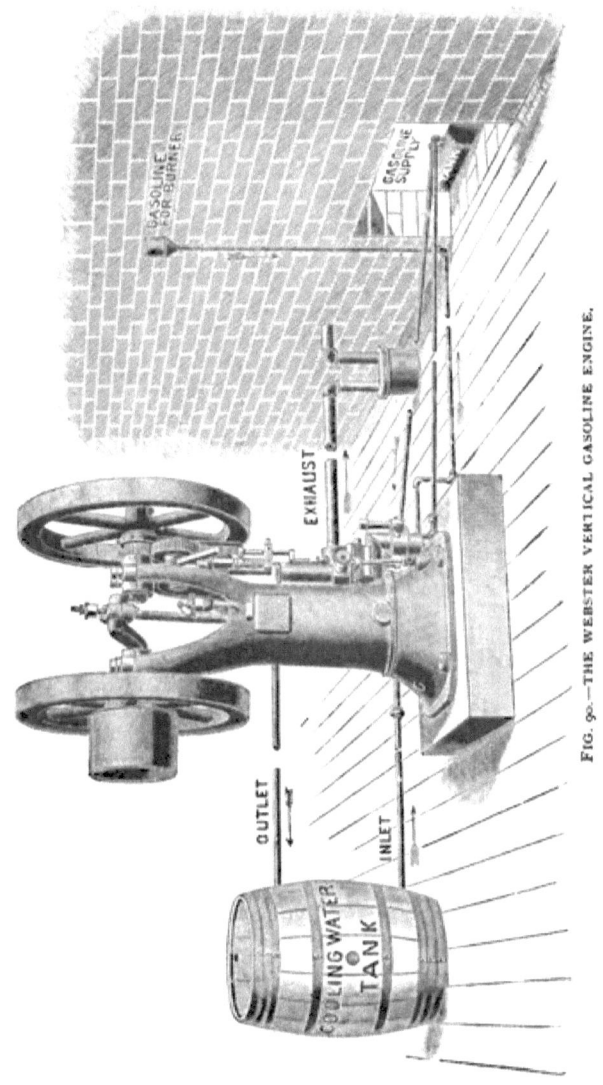

FIG. 90.—THE WEBSTER VERTICAL GASOLINE ENGINE.

tical gas and gasoline engine, with its connections with the gasoline supply, cooling tank, and muffler. The gasoline for the burner runs by gravity from a small tank on the wall. The vertical engines are made of 2 H.P. for power and pumping.

In Fig. 91 is represented the horizontal gasoline engine of this company. It is of the compression four-cycle type, with

FIG. 91.—THE WEBSTER GAS ENGINE.

poppet valves, tube igniter, gasoline pump, and regulating valves for both gasoline and air inlet, independent of the governor, which is of the centrifugal ball type, attached to the main shaft, and operates a regulating cam. The reducing gear from the main shaft, through a secondary shaft, operates the exhaust valve and gasoline pump through the lever across the front of the bed piece.

In operation, the air charge is drawn in through the pipe and regulator valve from the hollow bed piece and vaporizing chamber to the valve chest, the inlet valve opening by the suction of the piston.

When running light the governor shaft causes the exhaust valve to miss its lift, as also the gasoline pump to miss its stroke, and thus the gasoline supply is cut off until released by the governor. A small lever serves to open the exhaust valve and relieve the pressure in starting the engine.

A self-starting mechanism is furnished for the larger size engines, a novel and simple arrangement, consisting of a valve screwed into the top of the cylinder, in which is inserted an ordinary explosive match. By screwing the valve disc down to make tight, the head of the match comes in contact with the seat of the valve, which produces a flash and thus ignites the charge, which has been slightly compressed by turning back the fly-wheel with one hand, while with the other hand the operator turns the valve to its seat.

The sizes of engines made by this company are of 4, 6½, 10, 15, and 20 B.H.P., and adapted for the use of gas, natural gas, and gasoline.

The Springfield Gas Engine.

The engines of the Springfield Gas Engine Company are of the four-cycle compression type, adapted to the use of illuminating gas, natural gas, producer gas, gasoline gas, and gasoline fluid by injection.

The inlet and exhaust valves are of the poppet type, actuated by cams on a cross shaft over the cylinder head, the cross shaft being driven by a longitudinal shaft and two pairs of bevel gears.

The cams Nos. 18 and 19 on the cross shaft (Fig. 93) operate the inlet and exhaust valves by depression against internal pressure, the valves being also held to their seats by springs.

The governor is of the horizontal, centrifugal type, running free on the end of the cross shaft and driven by a small belt from the main shaft. Fig. 93 shows an end view of the engine as fitted for gas. An air valve No. 8 and the gas valve No. 35 are on a vertical spindle, which is operated by a cam, rotating with the cross shaft and controlled in its longitudinal

142   GAS, GASOLINE, AND OIL ENGINES.

FIG. 97.—THE SPRINGFIELD GAS ENGINE.

VARIOUS TYPES OF ENGINES AND MOTORS. 143

motion by the governor, making an off-and-on charge. The portion of air charge is fixed by the set of the air valve, and the proportion of the gas charge is regulated by adjustment of

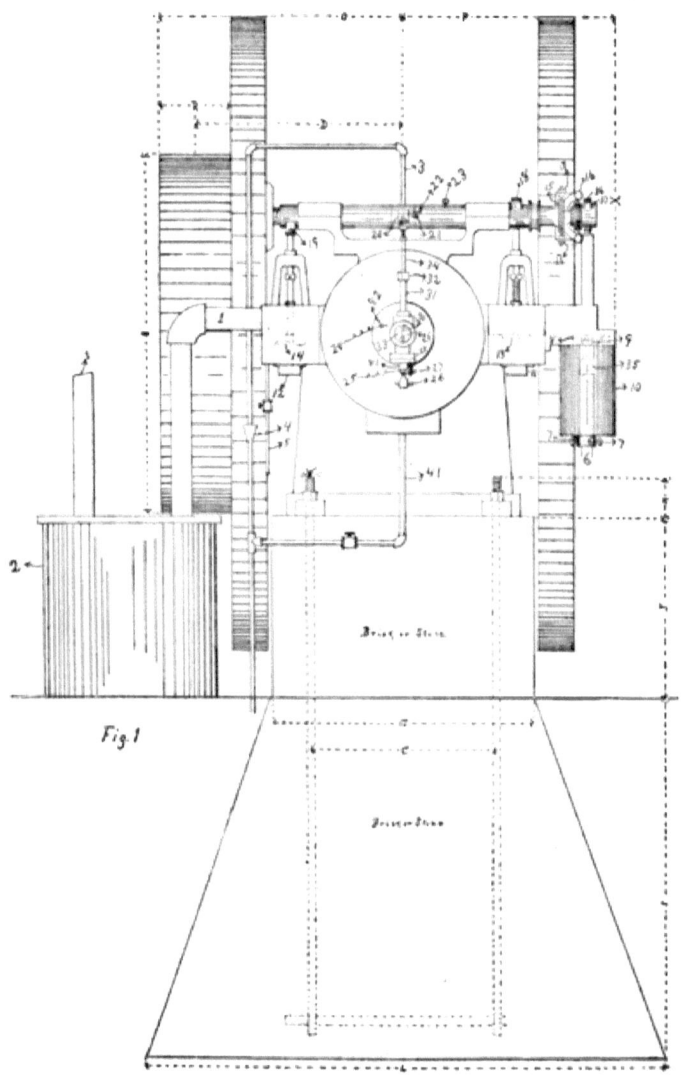

FIG. 91.—THE SPRINGFIELD GAS ENGINE—END VIEW.

the gas valve, which is set by raising or lowering the gas-inlet pipe No. 6 in the mixer No. 10 by means of the set-screws No. 7.

For the use of gasoline a small supply pump, driven from a cam on the longitudinal shaft, supplies the fluid to the injection plunger with an overflow to return the surplus to the gasoline tank.

Fig. 94 is a side view of the engine as arranged for control-

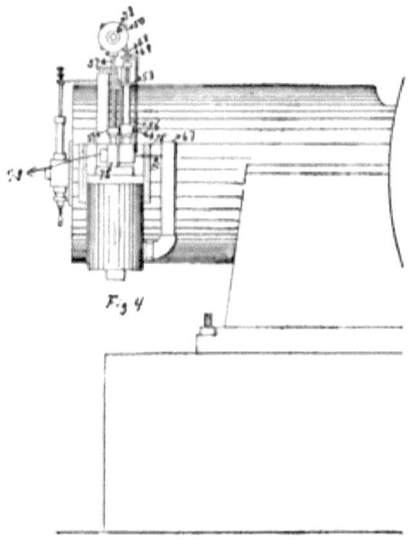

FIG. 94.—GASOLINE REGULATOR.

ling the fluid injection. The air-inlet pipe is attached to the side of the mixing tank; the gasoline pipe from the supply pump enters at No. 72. No. 56 is the injector plunger, and No. 57 the air-valve stem.

With a gravity feed the supply pump is dispensed with. Electric ignition is used. The device is embodied in a flanged chamber bolted to the head of the cylinder, as shown in Figs. 93 and 94, and the construction is detailed in Fig. 95. The upper electrode No. 34 vibrates as a current breaker, and is operated by a snap cam and spring lever at No. 20 in Fig. 93. The lower electrode is insulated and has a screw movement for adjusting the separation of the electrodes.

The battery connections are made on the head of the cylin-

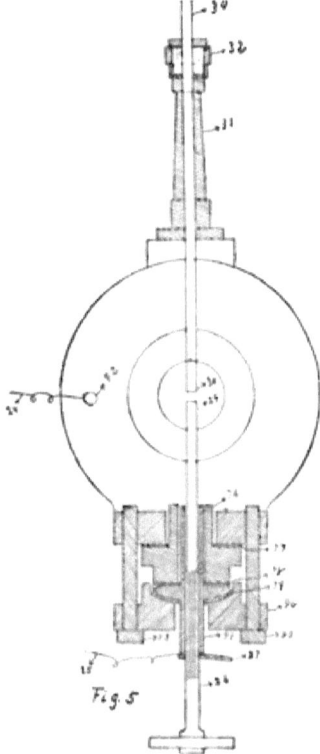

FIG. 95.—THE IGNITER.

der at the binding post 82, and to the insulated electrode at 25.

The battery plant consists of four (more or less) Edison-Lelande cells in series, a sparking-coil, and switch, as shown in Fig. 96. The sparking-coil is more fully described on page 75, in the chapter on ignition devices. The switch should always be turned off when the engine is not running, to save battery waste.

The Springfield Gas Engine Company builds eleven sizes of gas and gasoline engines, from 1 to 40 B.H.P. Full details for running these engines, with reference and key to the parts as figured, are given in their book of instructions.

The Foos Gas and Gasoline Engine.

The engines of the Foos Company are built in the horizontal and vertical style, and of 16 sizes from $2\frac{1}{2}$ to 100 B.H.P.

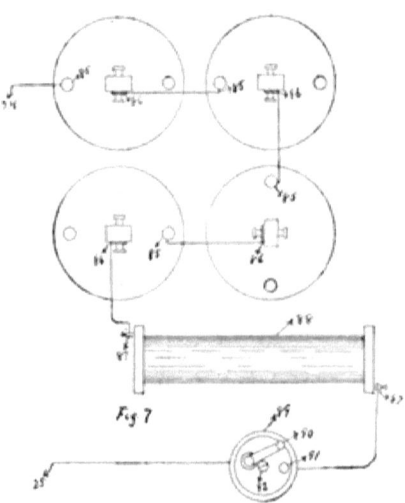

FIG. 96.—THE BATTERY.

They are all of the four-cycle compression type, with poppet valves. Fig. 97 represents the horizontal engine as connected for the use of gasoline.

The exhaust valve on the opposite side of the cylinder in the cut is lifted by a rock shaft and arms operated by a connecting-rod inside of the engine base, leading to a cam on the reducing-gear. The adjustable spring closes the exhaust valve. The regulation is made by mischarges of gas or gasoline by an interrupter device on the charge push-rod leading from a cam on the secondary gear. The governor L is of the

VARIOUS TYPES OF ENGINES AND MOTORS. 147

FIG. 97.—THE FOOS GAS AND GASOLINE ENGINE.

horizontal centrifugal type, driven by a band from a pulley on the main shaft. The movement of the governor operates a lever, which makes a hit-or-miss contact between the push rod and the pump rod, as may be traced by inspection of the cut (Fig. 97).

When gas is used, the pump is removed and a lever attachment made in place of the pump rod, which operates a gas

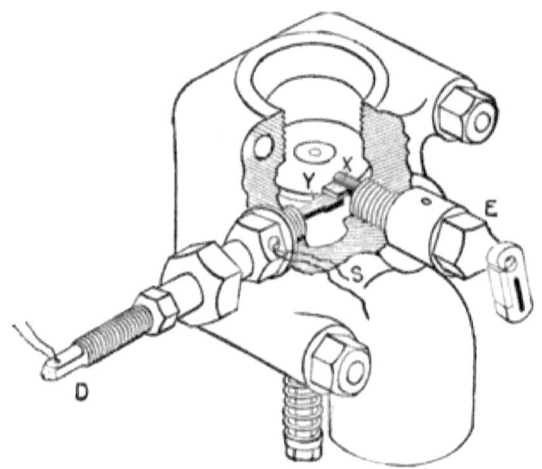

FIG. 98.—THE ELECTRODES.

valve for intermittent discharges into the air-inlet pipe, in the same manner that the gasoline injection is made, and controlled in the same way.

The charging and exploding chamber is shown at B (Fig. 97), and the details of its operation are shown in Fig. 98. The air is drawn in by the suction of the piston through the valve shown at X Y, the spindle of which passes through a subchamber connecting with the air pipe, and is regulated in its tension by a spiral spring and adjusting nut. The electrodes are shown at D and E, D being an insulated spring with its battery connection at D, and the opposite electrode is connected to the plug at S. The electrode E is revolved by the oscillating and sliding bar F, Fig. 97, one end of which is connected to an adjustable crank pin on the secondary gear, and the other to the crank of the electrode E. The slide pivot, as observed near the middle of the bar, enables the bar to transmit a circular motion to the electrode in an opposite direction from the motion of the pin on the secondary gear wheel. The time of sparking is regulated by moving the driving-pin in its circumferential position by turning the slotted plate K, in which the pin is set. The proper moment is at the end of the forward stroke of charge compression. A relief valve G is provided for relieving the pressure in the cylinder when turning over the fly-wheel for starting.

The speed of the engine may also be controlled by compressing or loosening the governor springs, by means of the nuts at each end of the springs.

The electric batteries are of the Edison-Lelande type in series.

The Dayton Gas and Gasoline Engine.

The engines of the Dayton Gas Engine and Manufacturing Company are built in the vertical and horizontal style, and also mounted as a portable engine on a wagon for agricultural purposes. They are of the four-cycle compression type, with the valve chamber on the top of the cylinder in the horizontal style, with poppet valves operated by straight-line push-rods from cams on the secondary shaft. The exhaust-valve rod with a back spring is on one side, and the admission valve with a positive cam motion and back spring is on the other side of the valve chamber, while between is the igniter rod, also operated by a cam—all having straight-line motions. The gas or gasoline valve is also operated by a rod and push-point, which is controlled by the governor.

The governor is of the horizontal, centrifugal style, mounted on the main shaft, adjusted by springs, and so arranged that the engine speed is regulated by hit-and-miss charges of gas or gasoline. The ignition is electric. The spark is produced by the end of the push-rod passing an insulated stem in the mixing-chamber, and made adjustable by a movable collar and handle between spiral springs. The handle on the igniter rod allows the electrodes to be readily cleaned

FIG. 99.—THE DAYTON ENGINE.

by vibrating the rod. The battery and sparking-coil is similar to those described with other engines. A match igniter for starting is also provided.

The Dayton is built in eleven sizes, from 2 to 50 H.P., and arranged for using natural and producer gas, illuminating gas, and gasoline.

The Victor Vapor Engine.

The engines of Thomas Kane & Co., are of the four-cycle compression type, with poppet valves, ignition by hot tubes or electric battery and double sparking-coil.

Fig. 100 is a view of the engine as fitted for gasoline with hot-tube igniter, with one fly-wheel off to show the arrangement of the valve gear. A cam on the secondary gear drives the push-rod lever of the exhaust valve, which is held back by a spiral spring. The governor is of the horizontal centrifugal type, revolving on the main shaft, and by overspeed carries

VARIOUS TYPES OF ENGINES AND MOTORS. 151

FIG. 100.—THE VICTOR VAPOR ENGINE AS ARRANGED FOR GASOLINE.

the roller of the push-rod lever on to the governor eccentric, holding the exhaust valve open.

The gasoline pump forces the gasoline into a small cup over the vaporizer, with an overflow back to the gasoline tank. The gasoline is fed to the vaporizer by a small valve and sight-feed cup, and comes in contact with the hot air drawn from the exhaust heater, which is a casing placed around the exhaust pipe and connected with the vaporizer by a side neck at the top of the vaporizer.

Thus the gasoline coming in contact with the hot air from the heater on extended surfaces inside of the vaporizer is completely vaporized and mixed with the air to saturation before it enters the admission valve, which opens by the suction of the piston.

Any accidental surplus of gasoline that may enter the vaporizer will drop into an extension of the vaporizer below the engine feed pipe, and flow back to the gasoline tank. An indexed regulating valve in the vapor pipe near the admission valve serves to regulate the flow of saturated vapor to the admission valve, where it is mixed with a further portion of air drawn in by the piston to make a proper explosive mixture.

The electric igniter is entered through the walls of the exhaust-valve chamber, which is directly connected with the inlet-valve chamber. It makes a double spark by a revolving mechanism driven from the secondary gear wheel and is adjustable, so that a spark takes place, one just before and one just after final compression—this being one of the peculiar features of this engine, from which a high efficiency is claimed; the other being the thin cylinder walls, as devised by Mr. Pennington.

In Fig. 101 the same engine is shown ready for gas connection, the operation of which is the same as for gasoline, as far as the valve action and regulation is concerned.

The sizes of the "Victor" are at present of 2, $3\frac{3}{4}$, and 5 B. H. P.

VARIOUS TYPES OF ENGINES AND MOTORS. 153

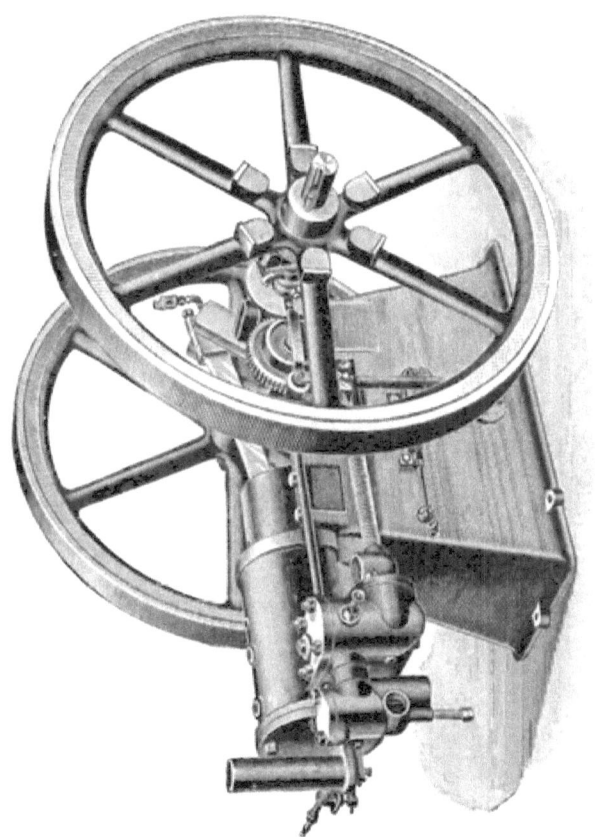

FIG. 101.—THE VICTOR VAPOR ENGINE AS ARRANGED FOR GAS.

The Wolverine Motor.

The engines of the Wolverine Motor Works are in the vertical style, for both stationary and marine power, as also for car-motor service. They are of the two-cycle and four-cycle compression type, with poppet and cylinder port valves. The stationary engines are for gas or gasoline of any grade from

FIG. 102.—THE JUNIOR STATIONARY.

.63 to .76 gravity. The marine engines use an injection of gasoline fluid into an air chamber, from which the vapor-and-air mixture is drawn into the closed crank chamber by the upward stroke of the piston.

The junior stationary engine (Fig. 102) is of the four-cycle class, taking its charge of gas or gasoline by the suction of the piston, compressing by the upward stroke, and exploding by a tube or electric igniter. The gasoline pump as shown in the cut is operated by a bell-crank lever and roller running on an eccentric on the secondary gear. The exhaust valve is operated from a cam also on the secondary gear. The speed is controlled by a simple governor, which consists of a single bar of steel, operating by the inertia of vibration. The junior is made with single cylinders from 1 to 6 H.P., and with double cylinders of 8 and 12 H.P.

In Fig. 103 is illustrated the two-cycle stationary motor. The charging-chamber and valve are located at the upper end of the cylinder, and the exhaust ports at the lower end of the

FIG. 103.—THE TWO-CYCLE STATIONARY.

stroke in the walls of the cylinder, and are uncovered by the piston at near the end of its down-stroke. The operation is as follows: The up-stroke of the piston draws a charge of air and gas into the crank chamber of engine, the down-stroke compresses the gas slightly in the base, and when the piston is near the end of the down-stroke a port is opened in the cylinder head which permits the compressed gas in the crank chamber to pass through a passage at the side of the cylinder through the open port of the cylinder head into the upper end of the cylinder. The next up-stroke of the piston compresses the explosive gas mixture, and when the piston is near the end of the up-stroke the charge of explosive gas is exploded by an electric spark, which drives the piston down. When the piston is near the end of the down-stroke it uncovers an annular port on the side of the cylinder which permits the exhaust to escape, and immediately after the exhaust port opens, the port in the cylinder head is opened, admitting a new charge, at the same time driving the balance of the exploded charge out of the exhaust port. This is repeated at every revolution.

FIG. 104.—THE MARINE ENGINE.

The stationary engines are made in sizes of ¾, 1, 2, and up to 12 H.P.

In Fig. 104 is illustrated the Wolverine single-cylinder marine engine. Its principles of action are the same as in the stationary engine, with the addition of a water-circulating pump driven from an eccentric, through a rock shaft; a reversing gear by which the motion of the engine is reversed, the same as with marine steam engines. It is reversed while running, and requires no handling of the fly-wheel for reversal. It is made in sizes of ¾, 1, 2, 4, and 6 H.P., with boat shaft and propeller complete.

VARIOUS TYPES OF ENGINES AND MOTORS.   157

In Figs. 105 and 106 are illustrated the double-cylinder marine engines of this company. The eccentric on this engine operates the water pump and exploders for both cylinders, both for the forward and backward gear.

FIG. 105.—THE WOLVERINE DOUBLE MARINE ENGINE—FRONT VIEW.

The generator is a pipe with an open fitting containing an air-check valve and a needle valve for adjusting the gasoline injection. The generator pipe leads to each crank-shaft chamber, with a light check to each opening to prevent back draught from one cylinder to the other by the alternate strokes of the pistons. The down-stroke of the piston opens an exhaust port through the walls of the cylinder, and at the same time compresses the explosive mixture that has been drawn in at the

FIG. 16.—THE WOLVERINE DOUBLE MARINE ENGINE—REAR VIEW.

previous up-stroke of the piston. A connection between the crank chamber and a valve chamber on top of the cylinder head allows the compressed air-and-vapor mixture to flow through a piston valve into the cylinder at the moment that the pressure is relieved by the exhaust. The return up-stroke compresses the gas mixture, which is exploded by the trip of the electric exploding-device. By a novel arrangement of sector and lever the engine is reversed.

Another device for reversing the propeller wheel, made by this company, is a double concentric shaft with a sleeve and lever, by which the longitudinal shifting of the centre shaft causes the blades to turn for stopping or backing.

The Fairbanks-Morse Gas Engine.

The engines of Fairbanks, Morse & Co. are all of the four-cycle compression type. The horizontal style is built in eleven sizes, from 3 to 70 B.H.P., and the vertical style of 2 B.H.P. The design of these engines, which is mostly based on the Caldwell-Charter patents, has a simplicity of construction in which the least number of moving parts has been a leading feature.

The valves are of the poppet type, the exhaust valve being operated by a direct line push-rod with a roller contact with the cam on the secondary gear; the roller being thrown on or off the cam by a bell-crank arm moved by the governor.

The governor is of the centrifugal type attached to the fly-wheel, counterbalanced by spiral springs and made adjustable by set nuts.

To the exhaust valve push-rod is attached an arm that operates the gas inlet-valve in connection with the air pipe extending from the base of the engine. The gas valve has an index valve to regulate the flow of gas.

A mixing-chamber in the head of the cylinder is insulated from the combustion chamber by an inlet-check valve, self-operating, held to its seat by a spring, and entirely enclosed within the mixing-chamber by the flanged projection from the cylinder head. This arrangement makes this a free-working valve and avoids leakage or undue friction.

Hot tube and electric ignition are used as preferred. The

electrodes are located in the head of the cylinder, with its sparking-device operated by the exhaust-valve push-rod through a second push-rod and arms.

FIG. 107.—THE FAIRBANKS-MORSE GAS ENGINE.

The engine as arranged for gas is shown in Fig. 107.

The gasoline engines (Figs. 108, 109, 110, and 111) of various sizes represent the arrangement for gasoline. They have a gasoline pump attached to the base of the engine directly under, and driven by a crank pin on the face of the exhaust eccentric. The pump drawing a supply from a tank placed in a safe place below the level of the pump, discharges into a small reservoir (P in Fig. 109, and also shown in the cylinder heads of Figs. 108 and 110), and overflows the surplus back to the tank. A small valve K in the reservoir P regu-

FIG. 108.—THE FAIRBANKS-MORSE GASOLINE ENGINE, 3 TO 5 H.P.

lates the flow of gasoline to the mixing-chamber. In the air pipe is a nozzle leading to the reservoir P, and the ingoing air draws from the nozzle the proper amount of gasoline to form a perfectly combustible mixture of gasoline and air.

Each suction of the engine draws up fresh gasoline from the reservoir P, and always the same quantity, as controlled by the supply or throttle valve K.

The self-starting devices are shown in Figs. 111 and 112, and consist of a small hand air-pump for medium-sized engines,

162   GAS, GASOLINE, AND OIL ENGINES.

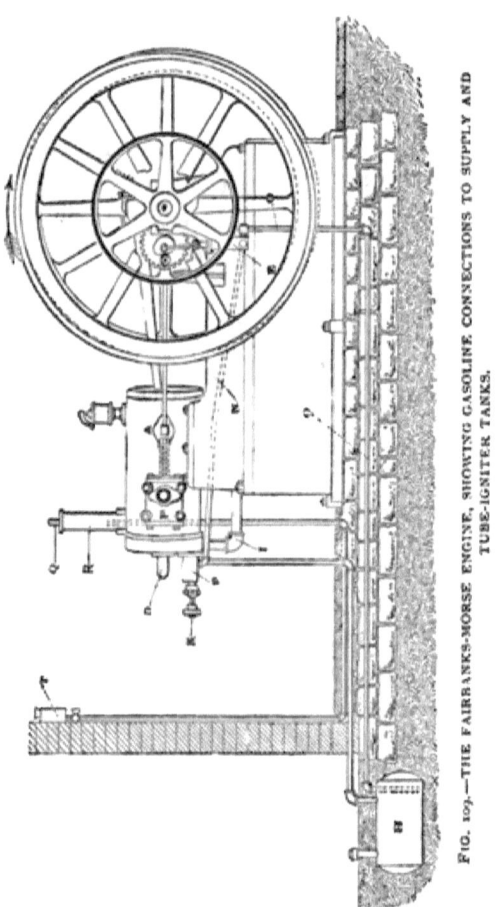

FIG. 109.—THE FAIRBANKS-MORSE ENGINE, SHOWING GASOLINE CONNECTIONS TO SUPPLY AND TUBE-IGNITER TANKS.

VARIOUS TYPES OF ENGINES AND MOTORS. 163

FIG. 110.—THE GASOLINE ENGINE, LARGER SIZE.

FIG. 111.—THE GASOLINE ENGINE, SHOWING THE SELF-STARTER CHARGING-PUMP.

VARIOUS TYPES OF ENGINES AND MOTORS. 165

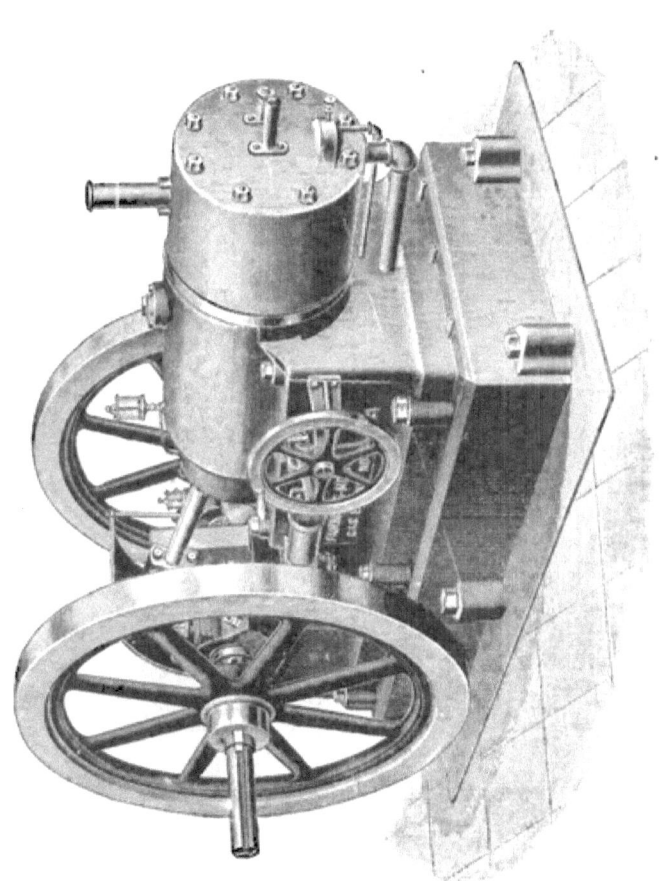

FIG. 112.—THE 50 H.P. GASOLINE ENGINE, WITH WHEEL AND CRANK PUMP FOR STARTING ENGINE.

FIG. 113.—THE VERTICAL ENGINE, SHOWING RATCHET CRANK FOR STARTING ENGINE.

Fig. 114.—THE VERTICAL GEARED ENGINE ON ONE BASE FOR PUMPING AND HOISTING.

and a hand crank pump on the larger size attached to the base of the engine. A small receptacle in the base of the pump is charged with gasoline of sufficient quantity for a single engine charge. The operation of the pump then charges the cylinder, and a match exploder fires the charge.

The small vertical engines of this company are illustrated in Figs. 113 and 114, for power and pumping purposes.

The bearings, crank, and valve gear are enclosed in the base and run in an oil bath, so that the piston and other moving parts are perfectly lubricated by the dash of the crank.

Fig. 113 shows the ratchet crank for starting the engine, and Fig. 114 shows the geared engine on one base as used for pumping or hoisting.

The Ruger Gas and Gasoline Engine.

The Ruger gas and gasoline engines are built in the vertical style, as in Fig. 115, of 1, 2½, 5, and 8 B.H.P.; and in the horizontal style, of 10, 15, 20, 25, 30, 35, and 50 B.H.P. They are of the four-cycle compression type; are arranged for gas, gasoline vapor or liquid, natural and producer gas. The gas engines have three poppet valves in two valve chambers, and the gasoline engines have only two poppet valves in one valve chamber.

Any of the valves can be quickly removed, cleaned, and replaced by the unscrewing of a plug. The adjustments are simple, and the ignition by hot tube or electric spark, as desired.

The governing is accomplished by controlling the exhaust valve; that is, holding it open when the speed is above the normal. The governor is located in the secondary gear, and by its centrifugal action retards the closing of the exhaust valve—thus relieving the piston from doing work by com-

VARIOUS TYPES OF ENGINES AND MOTORS. 169

FIG. 115.—THE RUGER VERTICAL GASOLINE ENGINE.

FIG. 116.—THE RUGER HORIZONTAL GAS ENGINE, 15 H.P.

pressing idle charges of air when the engine is running light.

The large sizes for electric lighting are built double, with impulse at every revolution of the shaft. For 30 H.P. and over, a self-starting device is provided. The gasoline pump is driven by an adjustable lever and rod operated from a cam on the reducing-gear.

FIG. 117.—THE RUGER, 10 H.P.

The pumping engines are vertical, and carry the pump and gear on the same base.

The igniting device is hot tube or electric, as preferred.

A special starting-device is furnished with the large engines.

The American Gas Engine.

The American Gas Engine Company have the control of the American patents of the Griffin gas engines, and of Dick Kerr & Co. of London, and Kilmarnock in Scotland. The Western Gas Construction Company are the manufacturers of these engines in all the patterns as made in Europe.

In Fig. 118 is illustrated their four-cycle compression engine, with poppet valves operated from a longitudinal cam shaft driven by spiral gear—the gas and air inlet entering

VARIOUS TYPES OF ENGINES AND MOTORS. 171

FIG. 118.—THE AMERICAN GAS ENGINE.

172 GAS, GASOLINE, AND OIL ENGINES.

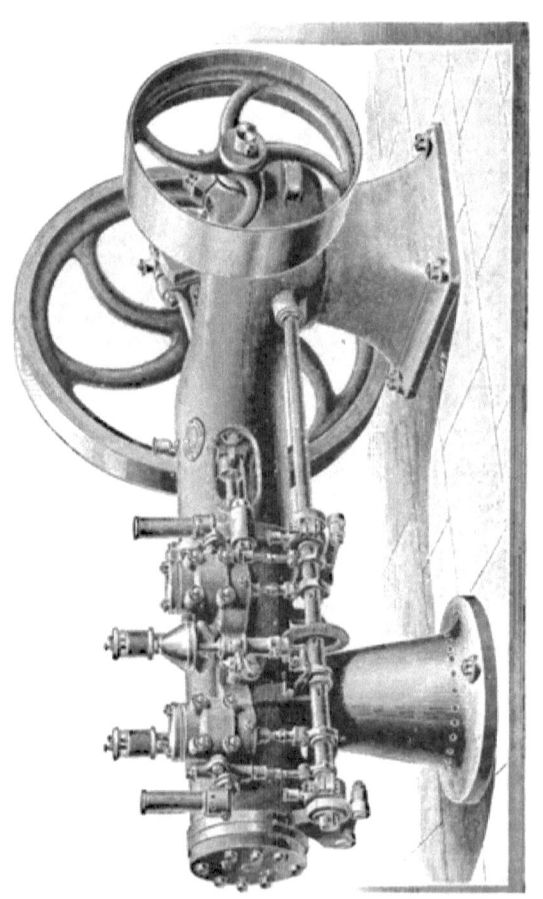

FIG. 119.—THE AMERICAN DOUBLE-ACTION GAS ENGINE.

through the cylinder head. The exhaust is on the opposite side of the cylinder; its valve is operated by a lever and roller from a cam on the valve-gear shaft.

In Fig. 119 is illustrated the double-acting engine of this company. It is essentially of the Griffin style as made in Europe, with an impulse on each side of the piston. The piston rod works through a stuffing-box in the front end of the cylinder, with the connecting-rod carried in a cross-head working in

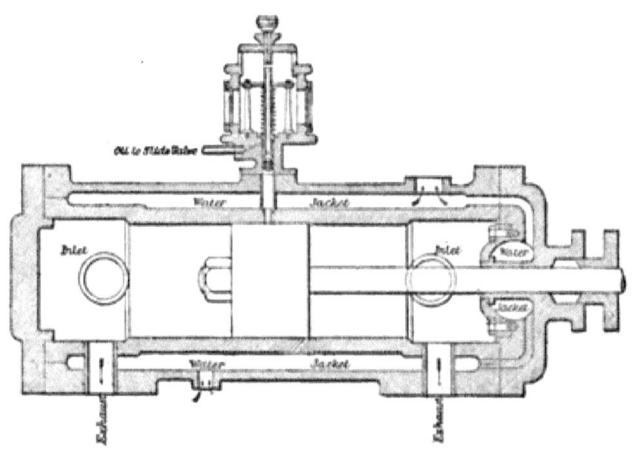

FIG. 120.—THE GRIFFIN DOUBLE-ACTING CYLINDER, TWO-CYCLE TYPE.

a slide frame, as in ordinary steam-engine practice. All the valves are of the poppet type, operated by cams on a single cam shaft, giving positive movement to every working part. Tube or electric ignition.

A ball governor, operated by bevel gear from the cam shaft, controls the gas inlet valve for both ends of the cylinder. The timing-valves are slide valves, also operated by cams on the cam shaft, and so arranged that the time of ignition can be adjusted and made uniform independent of the eccentricities of the hot tube.

In Fig. 120 is represented the construction of the cylinder

of the engine as made in England, showing the water-cooling jacket around the piston rod.

As a double-acting engine using the fourth stroke of the piston each way as an impulse stroke, it makes the action of the engine equivalent to a two-cycle type for steadiness of running. The single-acting engines are made in six sizes, from $1\frac{1}{2}$ to $11\frac{1}{4}$ B.H.P. The double-acting engines are made also in six sizes, from 4 to $18\frac{1}{2}$ B.H.P.

The Vreeland Gas Engine.

This engine is designed in the four-cycle compression type, with the principal exhaust through ports in the cylinder, un-

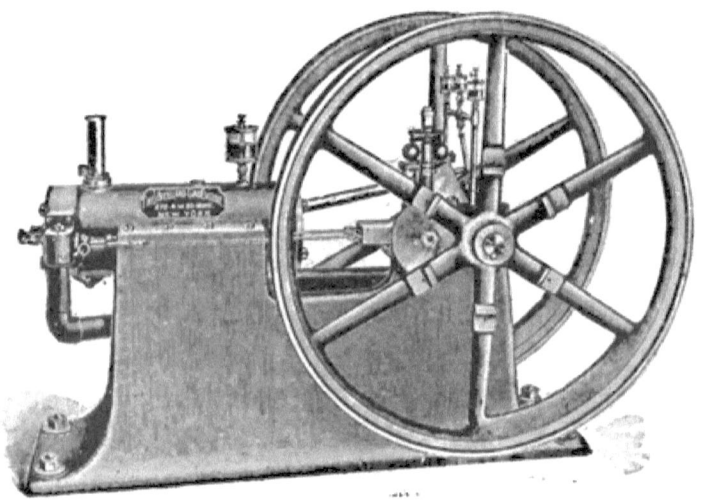

FIG. 121.—THE VREELAND GAS ENGINE.

covered by the piston at the end of the explosive stroke. It has also a supplementary exhaust valve in the head of the cylinder for completing the exhaust by the return stroke. The supplementary exhaust valve is operated by a lever across the cylinder head and a push-rod moved by a cam on the reducing gear.

The supplementary exhaust valve has a free communication by a pipe with the main exhaust. Both the cylinder and cylinder head have a water-cooling circulation. An independent push-rod from the gas-valve stem to a cam on the reducing-gear is controlled in its motion by the lateral movement of a roller, which is actuated through a bell-crank lever from the centrifugal ball governor. The governor is on a vertical spindle driven by a bevel gear attached to the reducing-gear—thus making a mischarge at the moment that the speed exceeds the normal adjustment of the governor.

Ignition is by hot tube on top of the combustion chamber.

A relief cock at mid-stroke facilitates easy starting. These engines are built in seven sizes, from 2 to 20 B.H.P.

The Backus Gas Engine.

The engines of the Backus Water Motor Company are built in the horizontal and vertical styles, as illustrated in Figs. 123 and 124. The horizontal engines are built in fifteen sizes,

FIG. 122.—THE BACKUS HORIZONTAL GAS ENGINE.

from 5 to 60 B.H.P. They are of the four-cycle compression type, with the principal exhaust ports in the side of the cylinder opened by the piston at the end of the impulse stroke. They have also a supplementary exhaust valve in the cylinder head, with its exhaust passage connecting with the main ex-

176   GAS, GASOLINE, AND OIL ENGINES.

haust. The exhaust push-rod is operated by an eccentric on the reducing-gear shaft, and carries a pendulum governor pivoted in the square box seen in the illustrations of the horizon-

FIG. 123.—THE BACKUS GAS ENGINE.

tal engines (Figs. 122 and 123). The push-blade of the governor is pivoted in the same box as the pendulum, with one end loosely locked in a Y-extension of the pendulum. The adjustment can be made while the engine is running, by a small screw seen in the front side of the small box, which compresses a spiral spring against a lug extending upward from the pendulum socket. The concave piston and cylinder head

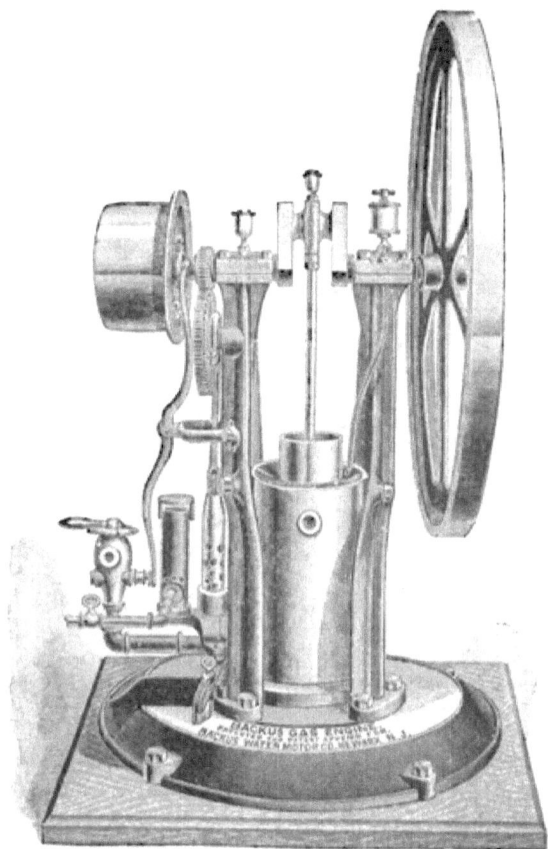

FIG. 124.—THE BACKUS VERTICAL GAS ENGINE.

are used in the Backus engines for the greatest volume in the combustion chamber with the least wall surface.

The Backus vertical engine is illustrated in Fig. 124, and a section in Fig. 125. The valves are of the poppet type. The exhaust valve has its motion controlled by a cam on the reducing-gear, while the gas valve is governed by a centrifugal governor in the pulley. The governing is by limiting or shutting off the gas, but the general regulation is made by an index valve. The gas inlet is through the air-inlet valve seat, so that when the engine stops the air valve closes the gas inlet by the

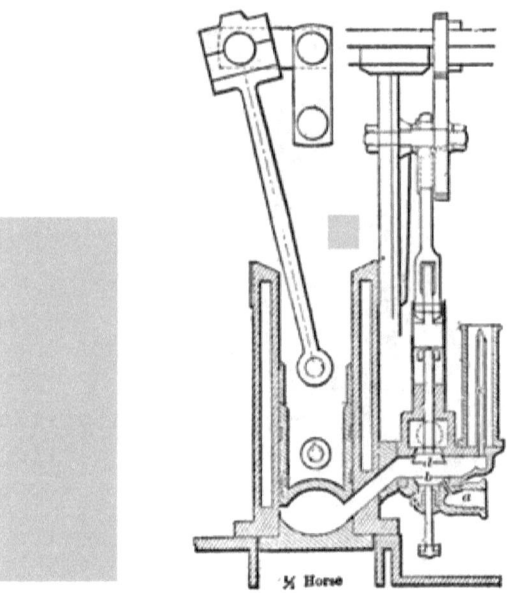

FIG. 125.—VERTICAL SECTION OF THE BACKUS GAS ENGINE.

action of its spiral spring, which is not shown. This is independent and automatic, and prevents the escape of gas by leaving the gas valve open.

The concave piston and cylinder head are shown in the cut; the gas inlet at a, combined gas-and-air valve at b, and the exhaust valve at d.

The Hartig Gas Engine.

The engines of the Hartig Standard Gas Engine Company are all made in the vertical style for gas or gasoline vapor, from a carburetter that gives a saturated air-vapor mixture, which is not explosive until a further admixture of air in the mixing-chamber of the engine completes its explosive quality.

FIG. 126.—THE HARTIG GAS ENGINE.

The engines are of the four-cycle compression type; ignition by hot porcelain tube or electric spark, and time igniter for the hot tube. The valves are of the poppet type. The exhaust valve is operated from a reducing-spur gear by crank pin, rod, and lever. The governor is of the centrifugal lever type, connected to a cam sleeve that has a circular motion by the movement of the balls, and a longitudinal motion by a spiral slot in

FIG. 127.—THE HARTIG PUMPING ENGINE.

the sleeve moving over a fixed pin in the main shaft. By this means the longitudinal movement of the sleeve rides the push-rod roller of the gas valve on to or off the cam, in such a way as to graduate the gas charges to meet the speed emergency.

The adjustment of the governor is made by spiral springs holding the balls in the position for normal speed.

VARIOUS TYPES OF ENGINES AND MOTORS. 181

The inlet-valve stem carries a double disc. The lower one is proportionally small for the gas passage, while the air is drawn in between the discs, the upper and larger valve discharging the mixture into the explosion chamber.

Fig. 126 illustrates the power engine, which is made in several sizes, from $\frac{1}{2}$ to 8 B.H.P.

Fig. 127 represents the pumping attachment operated from spur gear, all fixed complete on one base.

These engines as observed run on a consumption of from 18 cubic feet of gas in the larger sizes to 20 cubic feet in the smallest size per horse-power per hour. The pumping engines are of a capacity to force water to the highest city buildings.

The Allman Gas and Gasoline Engine.

The Allman engines are built in both the horizontal and vertical style. The horizontal engine (Fig. 128), in several

FIG. 128.—THE ALLMAN GAS AND GASOLINE ENGINE.

sizes from 2 to 15 B.H.P., is of the four-cycle compression type, mounted on a substantial iron base. The valves are of the poppet type, the exhaust valve being operated by a cam on the reducing-gear, and a roller disc on a lever actuating a second

FIG. 129.—THE ALLMAN VERTICAL.

lever at the valve stem through a connecting rod. The governor is a novel application in its adaptation to both governing and in balancing the crank motion.

The block shown on the hub of the fly-wheel (Fig. 128) is the frame plate of the governor, which supports a radial pin on which slides a rectangular block of steel, with angular

VARIOUS TYPES OF ENGINES AND MOTORS.   183

grooves on each side, in which the pins of a yoke lever slide by the centrifugal action of the steel block.

The other end of the yoke lever has also a yoke that straddles the sliding-sleeve on the main shaft, in which are pins en-

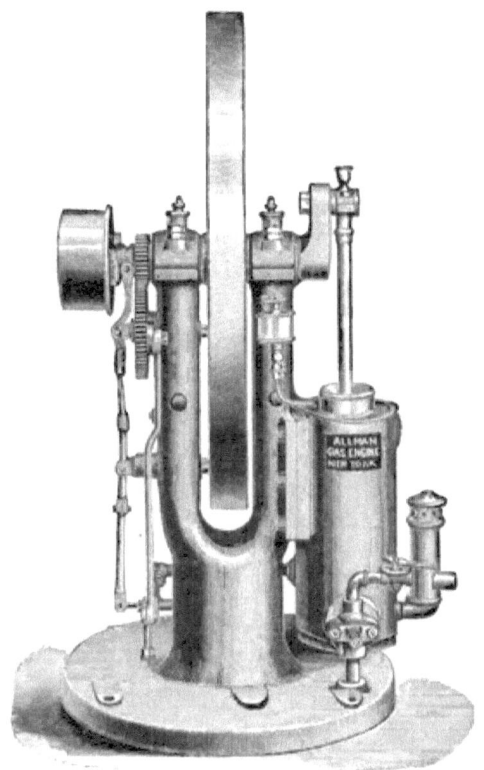

FIG. 130.—THE ALLMAN VERTICAL, ¾ H.P. ACTUAL.

tering a groove in the sleeve, and thus by the centrifugal action of the sliding steel block controls the movement of the sleeve in the direction of the axis of the shaft.

At the outer end of the radial pin, a spiral spring adjusted by a nut and check nut holds the steel sliding-block to the proper position at the normal speed of the engine. By the adjustment of the tension of the spring, the governor controls the engine at any desired speed.

A second groove in the sliding-sleeve operates a yoke lever and bell crank, touching the gas-valve stem with an adjusting screw—thus regulating the gas charge volume or cutting off as required.

The vertical engine, of this company (Fig. 129) are made on the same general principles as the horizontal type, and of 2, 3, and 4 B.H.P.

The governor on the vertical engine is of the horizontal, centrifugal ball type, with bell-crank movement of a sleeve on the main shaft—the governor being located in the pulley.

The lever, which is operated by a groove in the governor sleeve, extends down to and ending with a roller disc that rides on an adjustable wedge, resting on the arm of a rock shaft, the opposite arm of which lifts the gas-valve stem.

The range of travel of the push-roller on the wedge is limited by the governor, and thus makes a variable charge of gas.

The smallest size vertical of $\frac{3}{4}$ B.H.P. (Fig. 130) are constructed on the same general principles as the larger engines, but with a pedestal and base in one solid piece. The governing is in the same line as described for the larger vertical engines, but is applied to the exhaust valve, which is made to open partially or fully, or remain closed for regulating the speed—the wedge action for the exhaust valve being the same as for the gas charge in the other engines.

The Nash Gas Engine.

The Nash engines are built by the National Meter Company. They are of the vertical style, in nine sizes from $\frac{1}{8}$ to 10 H.P. with single cylinders; and in ten sizes from 10 to 200 H.P. with double and quadruple cylinders. The smaller engines are of the two-cycle compression type, taking an impulse at every revolution in each cylinder, thus making the action of

FIG. 132.—THE NASH VERTICAL ENGINE, SINGLE CYLINDER.

186   GAS, GASOLINE, AND OIL ENGINES.

FIG. 133.—THE NASH DOUBLE CYLINDER ENGINE, 10 TO 75 H.P., SPECIALTY FOR ELECTRIC LIGHTING.

the double-cylinder engines equivalent to the action of a single-cylinder steam engine or an impulse at each half-revolution of a single crank.

The double-cylinder engine (Fig. 133), the single cylinder

with double fly-wheel (Fig. 132), and the small single cylinder with one fly-wheel (Fig. 134) represent the general appearance of the engines of this company. They are all adapted for the

FIG. 134.—THE NASH, SMALL SIZES.

use of illuminating gas, gasoline, natural or producer gas. Ignition is by hot tube or the electric spark, as desired.

The larger engines have poppet valves, and are of the four-cycle compression type, and are now made in one-, two-, and four-cylinder vertical style, with reducing-gear and cam shaft, which operates the inlet and exhaust valve by direct-acting push-rods with back springs. The inlet-valve push-rods have bracket arms with pivoted push-blades that regulate the gas charge by the governor through a rock shaft and levers, which trip the push-blade contact for each gas-inlet valve.

This class of two- and four-cylinder engines is built in

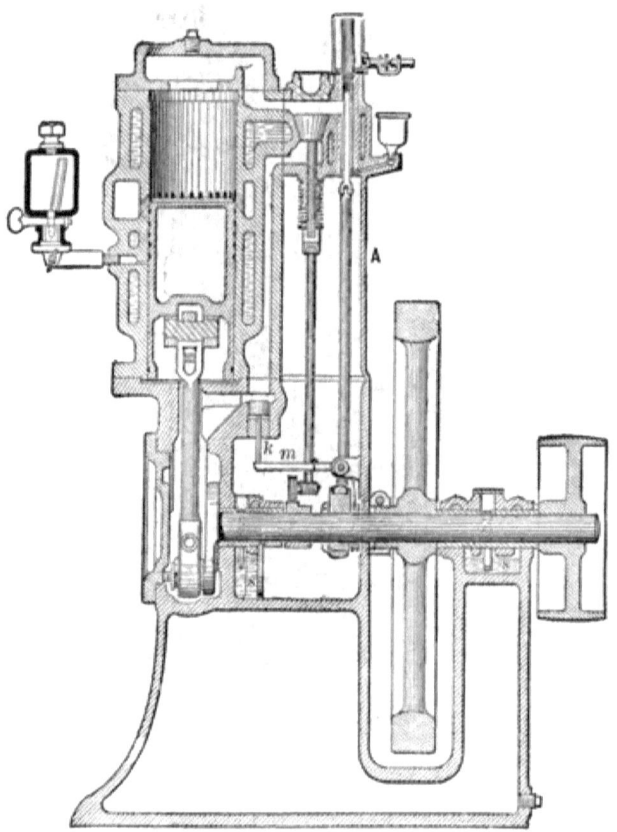

FIG. 135.—SIDE SECTION ELEVATION.

many sizes, ranging up to 200 B.H.P., with multipolar generators on the same base for electric lighting. Also combination pumping engines on a single base for deep wells; also combination engines and air compressors adapted to any required air pressure.

Some of the smaller Nash engines and the small pumping engines are provided with piston valves. In the two-cycle engines a combustion chamber is formed in the head of the cyl-

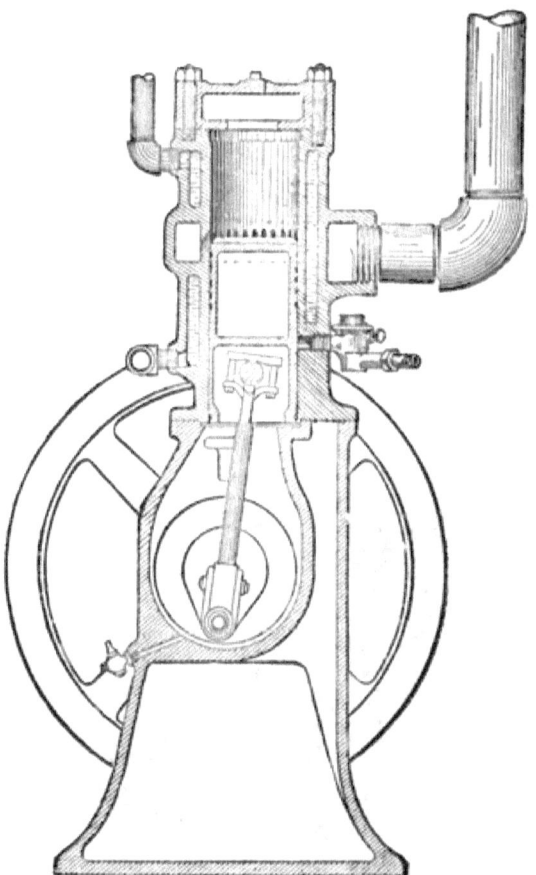

FIG. 136.—END SECTION ELEVATION.

inder, as seen in the sections (Figs. 135 and 136) into which the supply port and inlet valve opens. The lower end of the cylinder opens into a closed crank chamber, into which the gas-and-air mixture is drawn by the upward motion of the piston, through the mixing valve not shown. By the design of the mixing-valve the inflow of gas and air is adjusted partly by the relative proportions of the valve-seat openings. The flow of

FIG. 137.—GAS VALVE.

gas is further controlled by an independent index-gas valve (Fig. 137), so that the charge is always uniform in quality and density. By the downward motion of the piston the mixture

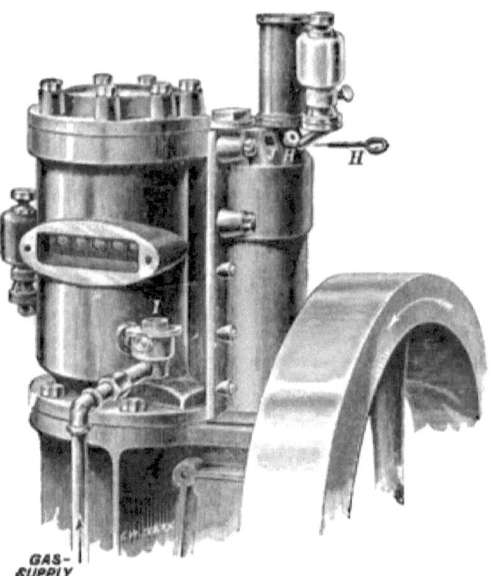

FIG. 138.—THE EXHAUST PORTS.

is compressed in the close crank case, and is supplied to the combustion chamber through a passage shown in Fig. 135, passing a valve, K, operated and controlled by the governor, for the purpose of varying the mixture charge to the needs for uniform engine speed. The larger inlet valve at the end of the passage is opened by a cam on the main shaft through a roller contact and push-rod, and closed by a spring.

The piston igniter is also a timing-valve, having a cavity,

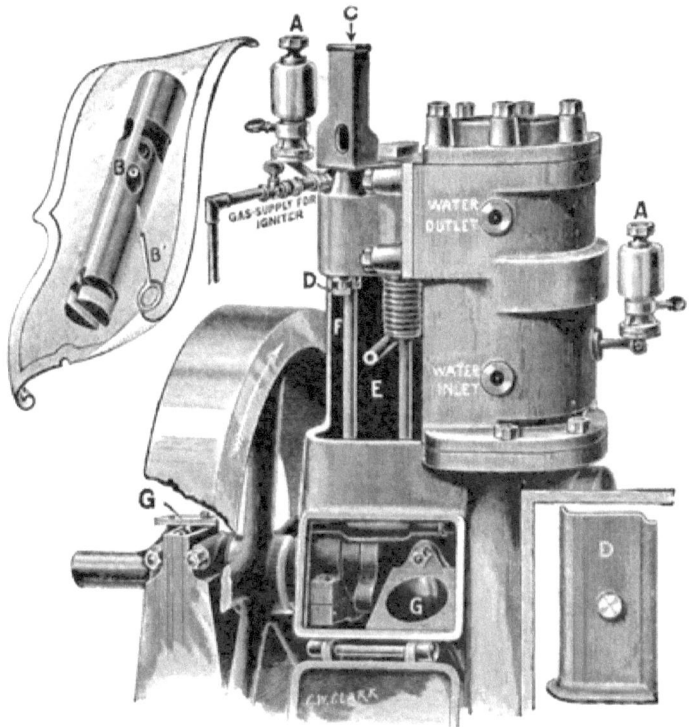

FIG. 139.—THE NASH VERTICAL WITH PISTON VALVE.

globular in shape, that receives its charge through a tangential opening that produces a vortical motion by which the gas and air are thoroughly mixed, and by a further movement of the piston the cavity is fired and the burning contents projected into the combustion chamber of the cylinder. It receives its motion from an eccentric on the shaft and a connecting rod.

These engines exhaust through ports in the cylinder at the

end of the piston stroke into an annular chamber on the outside of the cylinder wall. In Fig. 138 is shown the exhaust port chamber, cover off, with the ports in sight. This is one

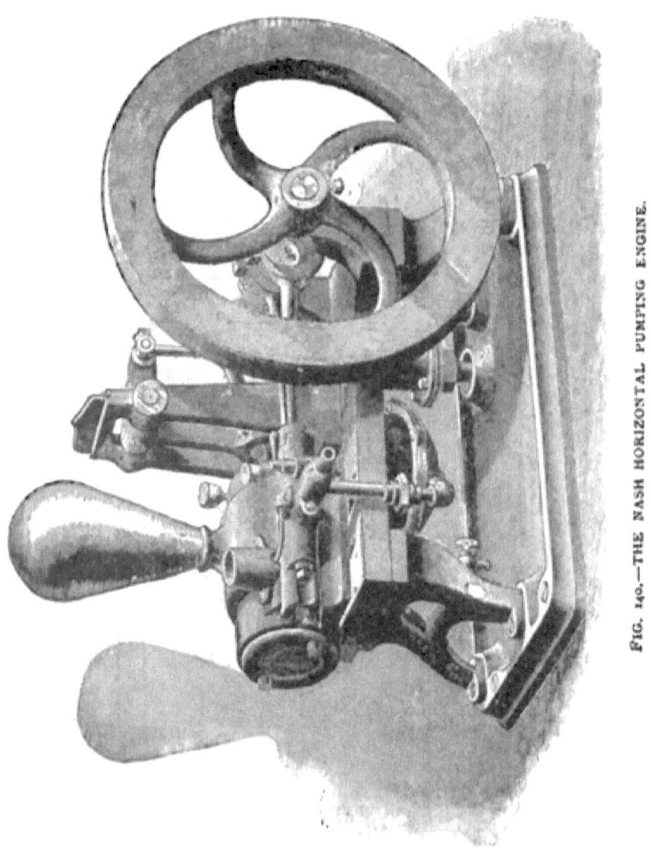

FIG. 140.—THE NASH HORIZONTAL PUMPING ENGINE.

of the earlier styles of the Nash engine with the gas-index valve opening through the side of the cylinder, with its inlet port uncovered during part of the upward stroke of the piston.

In Fig. 139 is shown the vertical engine, with the piston-ignition valve separate at the left of the engine cut. It is also shown in Fig. 32, in the chapter on ignition devices.

The Nash horizontal pumping engine (Fig. 140) is especially adapted for elevating water to the upper floors of buildings. It is of the two-cycle type, with piston gas and exhaust valves operated from eccentrics on the crank shaft. It is operated with either gas or gasoline.

The pump is located vertically within the engine frame, with a bell-crank lever above, and connecting rods to pump and engine pistons. This is the smallest engine made by this company, has a three-inch cylinder, four-inch stroke, and is equal to $\frac{4}{10}$ H.H.P. in water delivered, or 100 gallons 100 feet high per hour.

The Prouty Electro-Gasoline Engine.

The engines of The Prouty Company are built in the vertical style, from 5 H.P. upward. It is designed for stationary and road-wagon service, and for this last purpose the water-cooling arrangement is a departure from the practice in other engines, by the use of a small metal tank placed directly over the cylinder, as shown in the cut (Fig. 141). By the quick and direct circulation, the evaporation of the warm water and radiation of the tank surface are sufficient to keep the cylinder walls at the proper temperature.

The engines are of the four-cycle compression type, using poppet valves with electric ignition by contact points, operated from a cam on the reducing-gear shaft.

Primary or storage batteries are used. The governor is located on a disc attached to the reducing-shaft.

A gasoline pump, on the level with the tank at the left in the cut, is driven by a cam on the governor shaft and controlled by the governor. The gasoline is thus discharged in regulated quantity against the bottom of the intake valve; its opening is automatically closed, so that there is no possibility of spilling or discharge from the air inlet by the jarring or tipping of a wagon or carriage which the engine is driving. The pump has a positive throw controlled by the governor, which itself is not influenced by the jostling of a vehicle. The design of this engine was in view of its adaptation for driving road and traction wagons. It is also built for stationary power.

FIG. 141.—THE PROUTY ELECTRO-GASOLINE ENGINE.

A peculiar muffler made by this company gives a silent discharge of the exhaust so desirable in road and street motors.

Ignition by spark takes place in the inlet throat, between the valve chamber and cylinder, and at such time as to avoid the jar from sudden explosion at the exact end of the stroke of the piston.

The Lambert Gas and Gasoline Engine.

The engines built by the Lambert Gas and Gasoline Engine Company are all of the horizontal four-cycle type. They are

FIG. 142.—THE LAMBERT ENGINE, FRONT END VIEW.

scheduled in fifteen sizes, from 1 to 40 B.H.P. The valves are all of the poppet type and are operated by a secondary shaft and

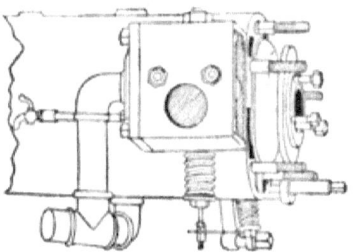

FIG. 143.—EXHAUST VALVE BOX, WATER HEAD OFF.

worm reducing-gear. The exhaust valve is opened by a lever across and under the end of the cylinder, the lever having a roller riding against a cam on the secondary shaft. The exhaust chamber (Fig. 143) has a water circulation through a jacket, and the cylinder head is also jacketed and connected, so that there can be no leak into the cylinder from the water circulation.

In Fig. 144 is shown the left side with the valve gear and

FIG. 144.—THE LAMBERT GAS AND GASOLINE ENGINE.

location of the governor, which is driven by a bevel gear on the secondary shaft.

In Fig. 145 is shown the detailed end view of the engine; the bell-crank lever that operated the gas-inlet valve from a cam on the secondary shaft, as also the sparking-cam o at the end of the shaft.

VARIOUS TYPES OF ENGINES AND MOTORS. 197

The spark-breaker and electrode are fixed on a small-eared flange bolted to the cylinder head, through which a rock shaft and insulated electrode pass. One arm of the rock shaft presses the electrode on the inside, while the outside arm is attached to a connecting rod, operated by the spring lever s and cam block k, which is adjustable. The amount of pres-

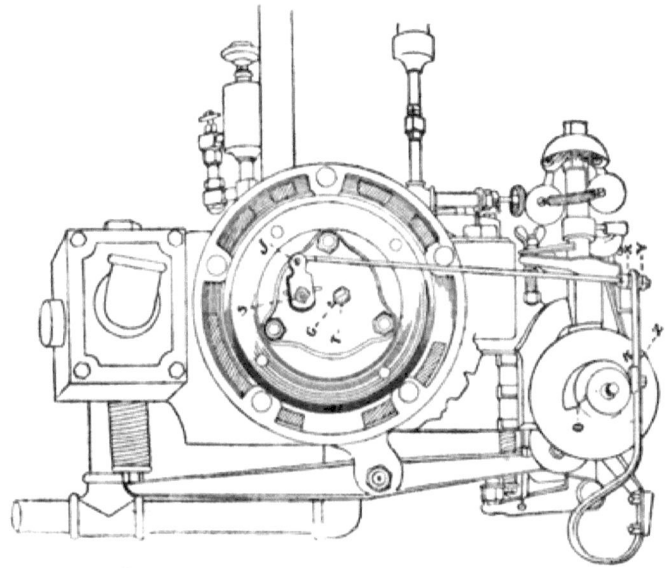

FIG. 145.—THE LAMBERT VALVE AND IGNITION GEAR.

sure of the inside arm is adjusted by the nuts x and y on the connecting rod.

In Fig. 146 is shown the electric battery, sparking-coil, and wiring, in which H and G are the binding posts on the valve chamber and insulated electrode. A relief cock is furnished for starting these engines.

In Fig. 147 is shown the gas regulator used with the Lambert engines—a most useful adjunct where the gas pressure is not uniform. A priming-cup for starting the gasoline engines and a gasoline pump operated by the cam shaft is not shown in the cuts.

198   GAS, GASOLINE, AND OIL ENGINES.

The "Leaflet" of directions issued by the Lambert Company is an excellent guide to the operator of a gas or gasoline

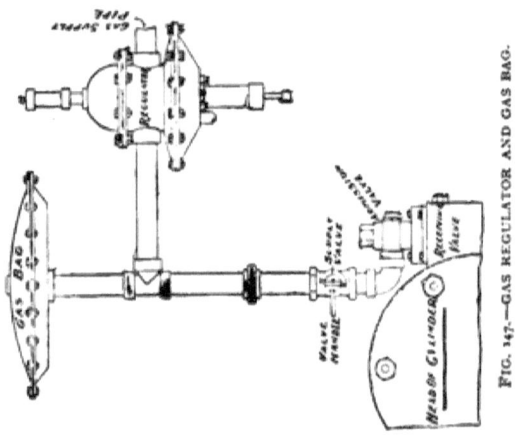

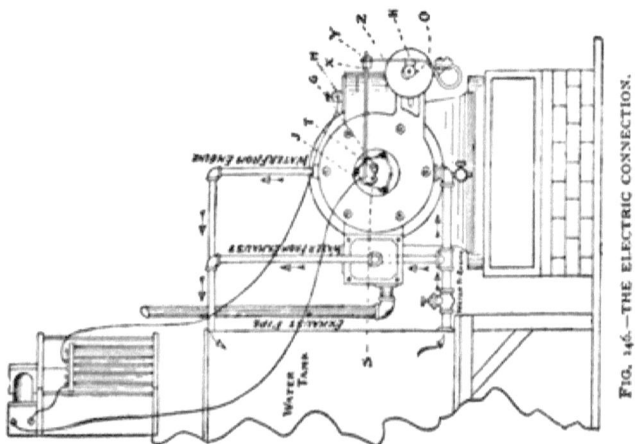

engine, and gives special directions for observing the internal action of the engine by the sounds to the ear.

VARIOUS TYPES OF ENGINES AND MOTORS. 199

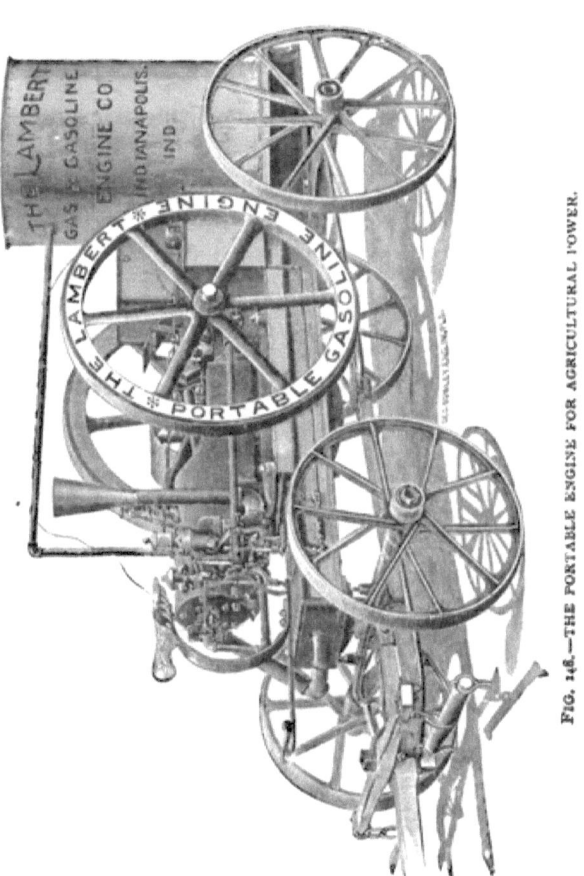

FIG. 148.—THE PORTABLE ENGINE FOR AGRICULTURAL POWER.

The Hicks Self-Starting Gas and Gasoline Engines.

In the engines of the Detroit Gas Engine Company a marked departure from the ordinary combination of cylinders for shortening the engine cycle has been made by placing two cylinders in tandem, by which an impulse is made for every revolution of the shaft. A piston rod, extending through a long sleeve between the cylinders, connects both pistons. The sleeve, which is the stuffing-box of the forward cylinder head, is packed by rings on the piston rod, which travel in the sleeve with the rod. The sleeve, being water-jacketed, avoids the difficulties heretofore met with piston rods running through ordinary stuffing-boxes and exposed to abnormal temperature in double-acting gas engines. With the Hicks engine the heated part of the piston rod is not a rubbing surface.

The valves are all of the vertical poppet style. The exhaust valves are operated through double-armed rock shafts centrally located under the cylinder, one arm of each moving in contact with alternating cams on the cam shaft.

The exhaust-valve chambers are water-jacketed. The cam shaft is driven with a reducing-worm gear, and dropped in its line position by a pair of spur gears for convenience of operating the valves. The inlet valves have also a positive motion directly from the cam shaft; as also the inlet valve for gas and gasoline, the mixture being made in a cross pipe between the nlet valves.

The gasoline pump is attached to the bed-piece, and is operated directly from a cam on the cam shaft through a bell crank with adjustment for pump throw. Electric ignition from batteries and spark coil by a break contact inside of the combustion chamber is used. An insulated platinum electrode with a rock shaft and tappet operated from a cam on the cam shaft through a pivoted lever for each cylinder, is the usual device for ignition. The governor is of the horizontal ball

VARIOUS TYPES OF ENGINES AND MOTORS. 201

FIG. 149.—THE HICKS COMPOUND CYLINDER GAS AND GASOLINE ENGINE, LEFT SIDE.

FIG. 150.—THE HICKS COMPOUND CYLINDER GAS AND GASOLINE ENGINE, RIGHT SIDE.

type, driven by spur-speed gear on the cam shaft, and through a push-rod varies the lift of the gas or gasoline valve, and thereby varies the charge.

The engines are equally well adapted for the use of coal gas, natural gas, producer gas, and gasoline. The regular sizes are at present eleven, from 3 to 55 B.H.P., with special power plants up to 300 H.P. The two-cycle effect of this engine gives it the uniform motion so desirable for driving electric generators for lighting purposes. The two views (Figs. 149 and 150) show the working details of this unique engine.

The American Motor.

This is a high-speed gas and gasoline motor made by the American Motor Company for stationary and marine service. It is as yet built in two sizes, of from 1 to 2 H.P. respectively, according to the fuel used, and of the style shown in Fig. 150; also as a twin engine with two cranks on one shaft of 2 to 4 H.P. Speed from 400 to 600 revolutions per minute. These engines are extremely light for their power, owing to the displacement of a water-jacket by the use of a coiled wire wrapping on the single-wall cylinder, which produces an extended air-cooling surface and dispenses with the use of water for cooling the cylinder.

These engines are of the four-cycle compression type, with but two valves, both with positive lift by push-rods and rollers with tension springs; the push rods are operated by cams, one on each side of the reducing-gear wheel. The gas or vapor enters through a graduating valve at the left in the cut, and the air through an opening under the inlet valve, also seen in the cut (Fig. 151). The insulated electrodes enter through the cylinder head, and are flashed by an induction or Ruhmkorff coil and dry battery. For stationary engines a governor is provided. Weight of the No. 1, 50 lbs., including fly-wheel without base; No. 2, 75 lbs., including fly-wheel without base

—being the lightest gas or gasoline engines in the trade for their power.

The adaptation of this engine in its portable form to the

FIG. 151.—THE AMERICAN MOTOR.

propulsion of small boats is a unique piece of mechanism. This adaptation is shown in Fig. 152, as applied to an ordinary rowboat of from 12 to 16 feet in length. By the hooked

VARIOUS TYPES OF ENGINES AND MOTORS. 205

FIG. 157.—THE AMERICAN MOTOR CO.'S PORTABLE BOAT MOTOR, WITH REVERSIBLE PROPELLER AND STEERING GEAR.

frame it is quickly dropped into place on the stern-board and clamped, the connection made with a carburetter at any convenient place in the boat with flexible tubing, and the boat is ready to start.

The motion of the vertical shaft inside the casing, seen at the water surface in the cut, is transferred to the propeller shaft by a bevel gear inside the rectangular case at the bottom. The blades of the propeller are rotated for stopping or backing by the movement of the grooved sleeve on the shaft casing and the bell crank, which transmits a reverse motion to the propeller blades. The lateral motion of the propeller and shaft for steering is made through the sector gear, and all the operations of steering, forward, stop, or backing, are made by two motions of the helm: a lateral motion for steering as usual for boats, and a vertical motion for changing the angle of the propeller blades. The cylinders of these little engines are $3\frac{1}{4}$ inches in diameter, four-inch stroke, and make from 400 to 600 revolutions per minute, with a boat speed of from six to eight miles per hour.

The Star Gas and Gasoline Engine.

These engines are built by the Star Gas Engine Company. They are of both horizontal and vertical style, as shown in Figs. 153 and 154.

The horizontal engine is built in eight sizes, from 1 to 25 B.H.P.

The vertical engines are built in one size, of 2 B.H.P.

The design is of the four-cycle compression type with poppet valves. The inlet valve serves also as a gas valve, having a broad seat with an annular slot connecting with the gas passage and gas-regulating or index valve.

The annular slot in the inlet-valve seat serves to thoroughly mix the gas and air at the moment of entering the combustion chamber.

A vertical ball governor driven by a bevel gear on the side

VARIOUS TYPES OF ENGINES AND MOTORS. 207

FIG. 153.—THE STAR GAS ENGINE.

FIG. 154.—THE VERTICAL STAR.

of the reducing-spur gear operates through a bell crank, the lateral movement of a disc revolving on a pin fixed in the gas-and-air-valve push-rod for making a graduating or hit-and-miss charge. An arm on the push-rod is adjustable for regulating the throw of the valve.

Some of the engines of this company are controlled by a pendulum governor, working on the inertia principle and using no springs. Ignition is by hot tube, which is placed on the top of the cylinder in the horizontal engine, leaving the cylinder head free to be removed without disturbing the attachments. In the vertical engine the igniter is fastened to the cylinder head.

The Daimler Motors.

The Daimler Motor Company, manufacturers of stationary gas, gasoline, and kerosene motors, and gasoline motors for boats, carriages, street-railway cars, fire engines, and portable electric lighting, are the sole owners of the United States and Canadian patents of Gottlieb Daimler, of Canstadt, Germany.

Their motors are all of the four-cycle compression type, following the principles formulated by M. Beau de Rochas, and carried out practically by Otto and Daimler in Germany, and now made by this company with many improvements derived from experience. All the valves are of the poppet style, closing automatically with springs. In the earlier engines and those of the duplex style with a single crank, the governing was made by a miss in the push-rod blade on the exhaust-valve stem by which the exhaust valve remained closed through a single cycle or more, as required by the action of the governor —the governor being of the horizontal centrifugal style, located in the pulley on the main shaft or in the fly-wheel when an outside fly-wheel is used.

The operation of the governor is transferred through a grooved sleeve to the lateral arm of a bell-crank push-blade on the push-rod of each cylinder, by a vertical pivoted lever carrying a stop-block, which is thrown out and into contact with the arm of the bell-crank push-blades, and makes a miss-opening of the exhaust valve, as shown in the duplex motor (Fig.

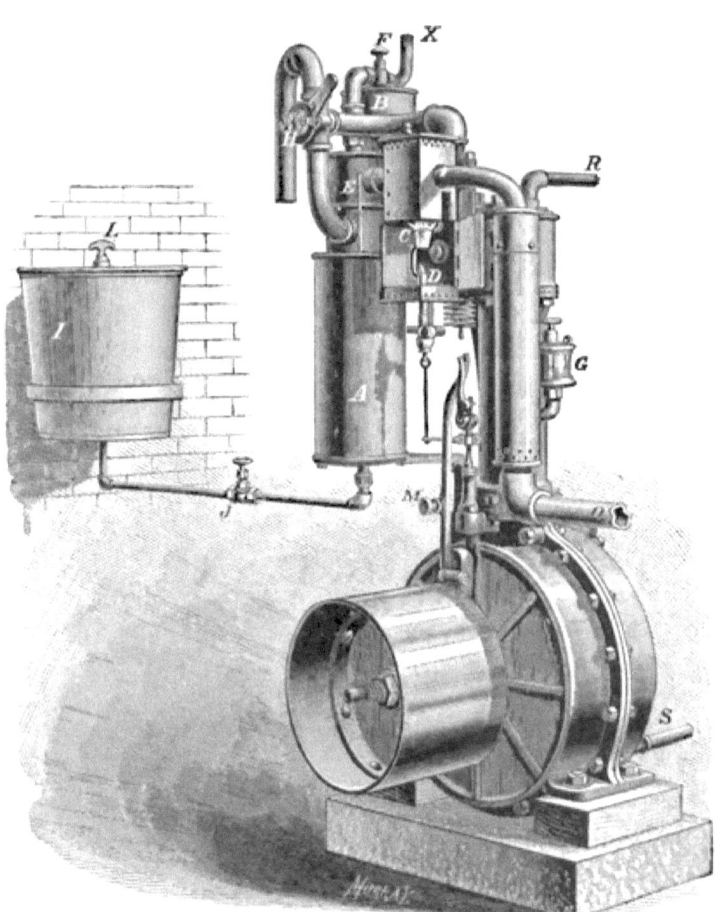

FIG. 155.—THE DAIMLER GASOLINE ENGINE, WITH CARBURETTER AND TANK READY FOR RUNNING.

A, carburetter; B, supply reservoir for burner, regulated by the valve F; D, the burner; C, the platinum ignition tube; H, the regulating valve for the mixture from the carburetter and free air; I, gasoline supply tank for carburetter; O, exhaust pipe, with air jacket for supplying warm air to the carburetter.

210 GAS, GASOLINE, AND OIL ENGINES.

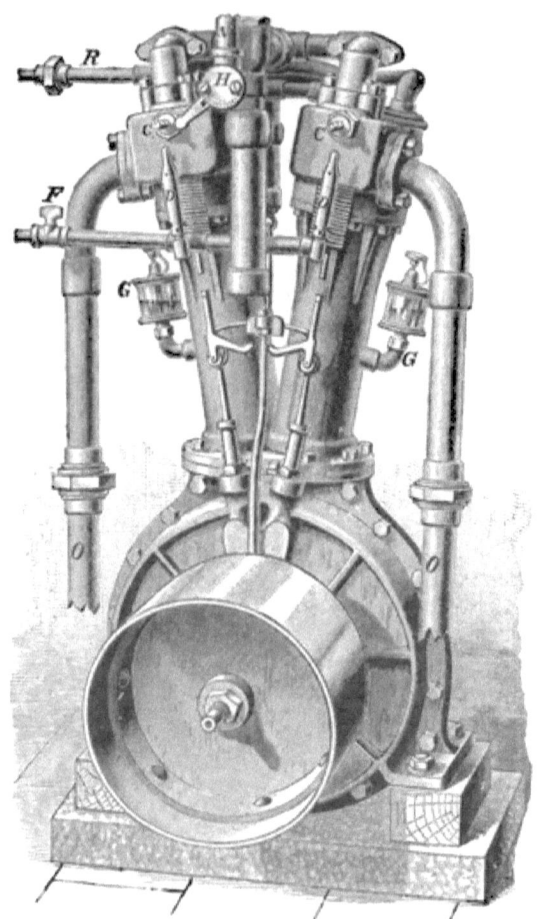

FIG. 156.—THE DAIMLER TWO-CYLINDER GAS ENGINE.

Showing the burners D, D; platinum igniters C, C; the gas flow pipe R; and regulating valve H; and the exhaust valve-gear with regulating stop-block and governor rod operated by the governor located in the pulley; N, the free-air inlet; F, the regulating cock for the Bunsen burners.

156), and also in the single-cylinder motor (Fig. 155). By this arrangement the movement of the piston, with the exhaust valve closed, simply compresses and recompresses the burned gases, and allowing no fresh charge to enter the cylinder until by the return to normal speed the governor allows the push-blades to act on the exhaust-valve spindle.

The ingenious mechanism by which the alternating motion of the valves is secured without the use of gearing for both the double and single cylinders is worthy of notice. By this arrangement the reducing-gear and its noise have a substitute in the eccentric double continuous groove, in which sliding-pin blocks perform the operation of a single eccentric for each cylinder. The pin-blocks and push-rods being off from a radial line, allow the blocks to cross successively the intersection of the eccentric groove.

In the new style of motors of this company the adaptation to the most ready fuel to be found in all parts of the world (kerosene), has made this style of motor a most desirable one for the foreign trade as well as a most economical one for home use.

Fig. 158 represents one of the new style small motors with enclosing case for the crank and connecting rod, while the outside reducing-gear and governor is enclosed within the area of the fly-wheel, making a most convenient and compact motor for all purposes of power.

In the kerosene motor the oil is vaporized by the heat of the exhaust by means of a jacketed evaporator, which only holds a moderate charge and is fed from a storage tank at a safe distance.

The single-cylinder motors are made from 1 to 12 B.H.P., and the double-cylinder motors from 4 to 24 H.P. The four-cylinder motors are made up to 48 H.P.

These motors have been adapted to marine propulsion to a large extent. Fig. 160 represents a 4 H.P. marine motor of the two-cylinder style on single crank, making the combina-

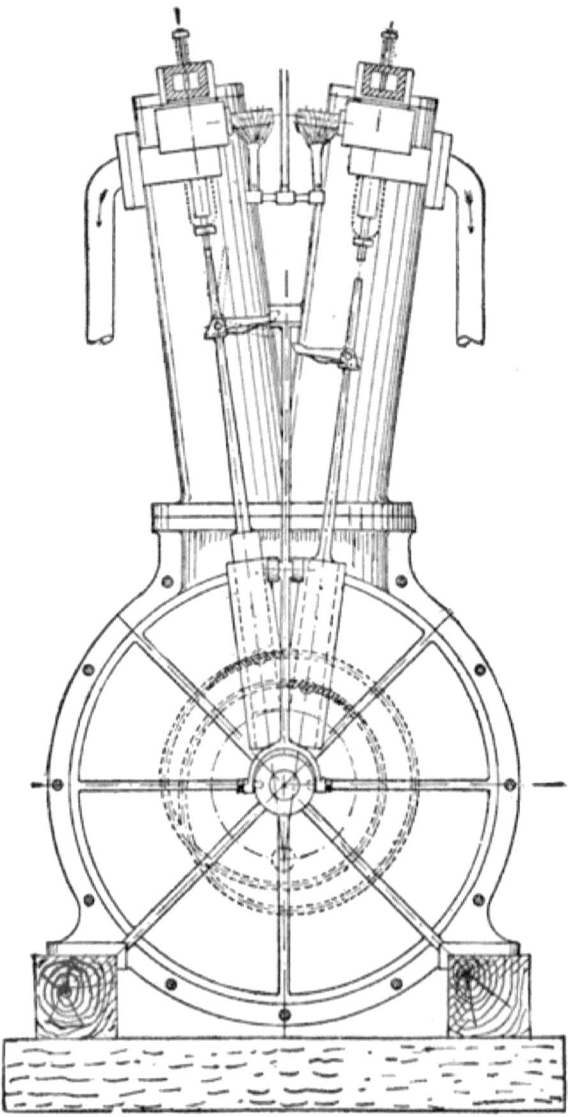

FIG. 157.—SIDE ELEVATION OF MOTOR.

Showing grooves in face of fly-wheel that control the exhaust valves for alternating the impulse in each cylinder.

VARIOUS TYPES OF ENGINES AND MOTORS. 213

FIG. 158.—THE NEW DAIMLER GASOLINE ENGINE.

tion equivalent to a two-cycle engine. With this engine the governor controls the speed with the variable load caused by stopping, slowing, or reversing the propeller wheel—all of these movements being controlled by the lever shown in the

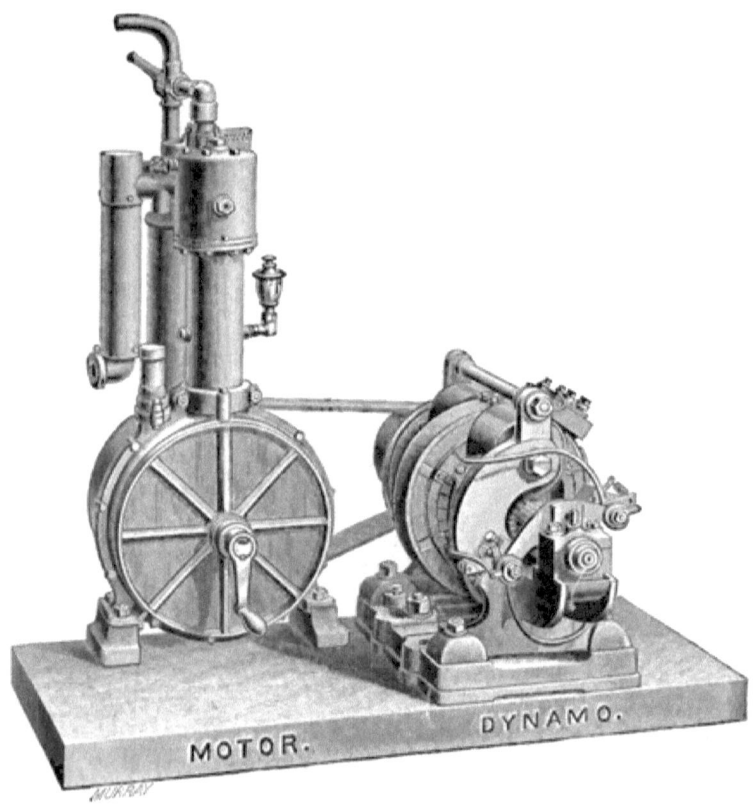

FIG. 259.—THE SINGLE-CYLINDER MOTOR AND ELECTRIC GENERATOR.

Also with two and four cylinders on one shaft for general electric lighting plants, giving a uniform and steady light, from 25 to 600 incandescent lamps.

cut. The first back pull of the lever eases the friction-clutch, which is the driving connection of the engine with the wheel shaft. A further pull unships the driving-clutch, and a still further pull puts the bevel-friction gear in contact for reversing

FIG. 160.—THE FOUR HORSE-POWER MARINE MOTOR.

216 GAS, GASOLINE, AND OIL ENGINES.

FIG. 161.—THE 42-FOOT FULL-CABIN DAIMLER MOTOR LAUNCH, 12 H.P.

VARIOUS TYPES OF ENGINES AND MOTORS. 217

FIG. 164.—THE DAIMLER RAILWAY INSPECTION CAR.

the wheel. The marine motors are all made for gasoline fuel.

In Fig. 161 is represented one of the cabined yachts of this company. The gasoline is stored in a copper tank in the bow of the boat, sufficient for a 60 to 150 hour run.

The 16- and 18-foot boats have 1 H.P. motor; 21-foot boat a

FIG. 163.—THE DAIMLER MOTOR BUGGY OR QUADRICYCLE.

A 1 H.P. motor and gear is located beneath the seat with the machinery so arranged that a single lever performs all the functions of starting, stopping, and steering the vehicle. The forward wheels turn each on its own forked axis and are linked together with the steering lever, which operates for steering in a horizontal direction and for starting and stopping in a vertical direction. Their four-seat carriage is of somewhat heavier build with a 4 H.P. motor.

2 H.P. motor; 25-foot boat a 4 H.P. motor; a 30-foot boat, 7 H.P., etc. The larger boats, up to 50 feet in length, have the entire control of the engine from the pilot house. The company are prepared to build and equip yachts up to 100 feet in length and with all the modern finish. The horseless carriages, buggies, inspection cars, street-railway cars, and fire-

engines are now scheduled in the manufacture of this company, which is associated with companies of similar name in London, Paris, and Canstadt, Germany.

In Fig. 162 is illustrated one of the railway inspection cars of this company, made to carry two inspectors and the motor driver. The motor is located behind the wheels, vertically, and belted to a pair of pulleys on the main shaft for two speeds. The change speed, stop, and start are made by friction-clutches, operated by one lever handle; the other lever is for the brake.

In Fig. 163 is shown the Daimler motor buggy or quadricycle. A 1 H.P. motor and gear is located beneath the seat, with the machinery so arranged that a single lever performs all the functions of starting, stopping, and steering the vehicle. The forward wheels turn each on its own forked axis, and are linked together with the steering lever, which operates for steering in a horizontal direction, and for starting and stopping in a vertical direction.

Their four-seat carriage is of somewhat heavier build, with a 4 H.P. motor.

The Olds Gas and Gasoline Engine.

We illustrate in Fig. 164 the latest design of gas and gasoline engines built by P. F. Olds & Son. These engines are of the four-cycle compression type, with poppet valves larger than the usual size to facilitate the exhaust and charge, and to avoid the counterpressures usual with small-sized valves.

The valve gear is a simple eccentric on the main shaft connected by a rod to a slide bar, moving in a bracketed box at the side of the cylinder. The slide bar carries a revolving alternating or toothed wheel, the alternating motion of which is governed by a pendulum swinging upon a concentric pivot.

The ratchet and toothed wheel are pivoted to the slide, and the teeth become push-pins to the spindle of the exhaust valve, and are made to open the exhaust regularly at normal speed and make a miss by throwing the notch in the wheel opposite

FIG. 164.—THE OLDS GAS AND VAPOR ENGINE.

VARIOUS TYPES OF ENGINES AND MOTORS. 221

the spindle when the speed is above the normal. By throwing out the pawl which operates the alternating wheel, compression will be omitted by the open exhaust, and the engine can be

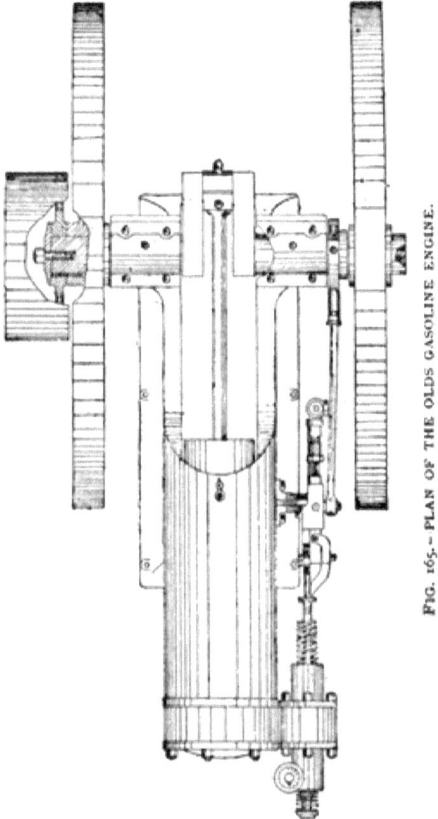

FIG. 165.— PLAN OF THE OLDS GASOLINE ENGINE.

easily turned to any point for starting without the resistance of compression.

The inlet valve is opposite and in line with the exhaust valve, and is opened by the suction of the piston. The vaporizing chamber for gasoline is in front of the cylinder head, and receives near its bottom the air pipe from the engine-bed frame.

When running with gasoline, a small pump is operated by

the eccentric rod, which supplies a small reservoir over the inlet valve, arranged so that the surplus runs back to the reservoir below the level of the pump, thus avoiding the possibility of accidental overflow of gasoline. On the top of the reservoir is a sight glass that shows the flow of the gasoline, with a set valve to regulate the feed to the mixing-chamber, where it is atomized by the inrush of air to the cylinder during the charging stroke.

The igniter is by hot tube or electric, preferably a hot tube, with some special improvements that make this style of ignition very desirable. The igniters are not shown in the cut, but occupy the place of a plug seen on top of the valve chamber.

This company also makes a vertical engine on the same principles as the horizontal one, in sizes of from 1 to 5 H.P. Their horizontal engines are made in five sizes, from 7 to 50 B.H.P. Also double-cylinder launch engines and launches—2 H.P. for 18- and 20-foot launches, 4 H.P. for 25-foot, and 8 H.P. for 35-foot launches. In these launch motors the gasoline for a day's run is stored in an iron receptacle at the motor, thus avoiding all danger from pipes and separate tank leakage.

In these boats the engine is not required to be set exactly in line with the propeller shaft. A reversing friction-clutch is used with a flexible shaft connection, so that the setting of the engine and shaft in any boat is an easy matter. The cooling water from the cylinders is discharged through the exhaust pipe, which is a rubber hose passing out at the stern. By this arrangement the rubber exhaust pipe is kept cool, and its flexibility makes a silent exhaust.

The Weber Gas and Gasoline Engine.

The engines of the Weber Gas and Gasoline Engine Company are of the four-cycle compression type, with poppet valves operated by direct push-rods and cams on the reducing-gear, which is enclosed with the governor in an iron box, partly filled with oil, which insures perfect lubrication of the gear and keeps out dust. The horizontal styles are made in eight sizes, of 3 to 15 B.H.P., as shown in Fig. 166; and in ten sizes, from 18 to 100 H.P., of the style as shown in Fig. 170

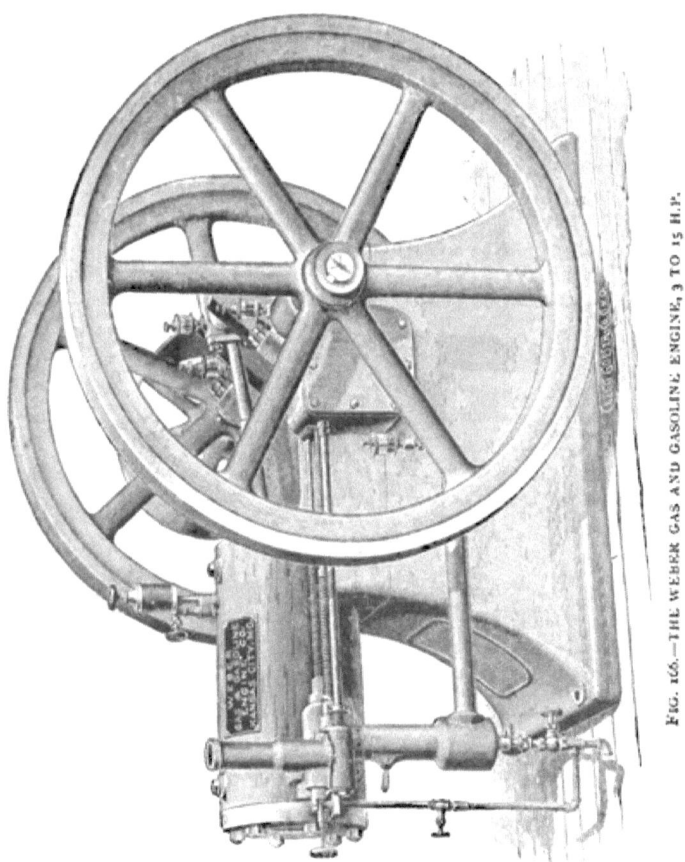

FIG. 166.—THE WEBER GAS AND GASOLINE ENGINE, 3 TO 15 H.P.

They also build a one size vertical engine, of 2 B.H.P., for pumping water, running ventilating fans and printing presses, etc., as shown in Fig. 168. The illustration (Fig. 169) represents a self-contained gasoline engine hoister, of 10 B.H.P.—a reliable and compact machine, designed to meet the wants of

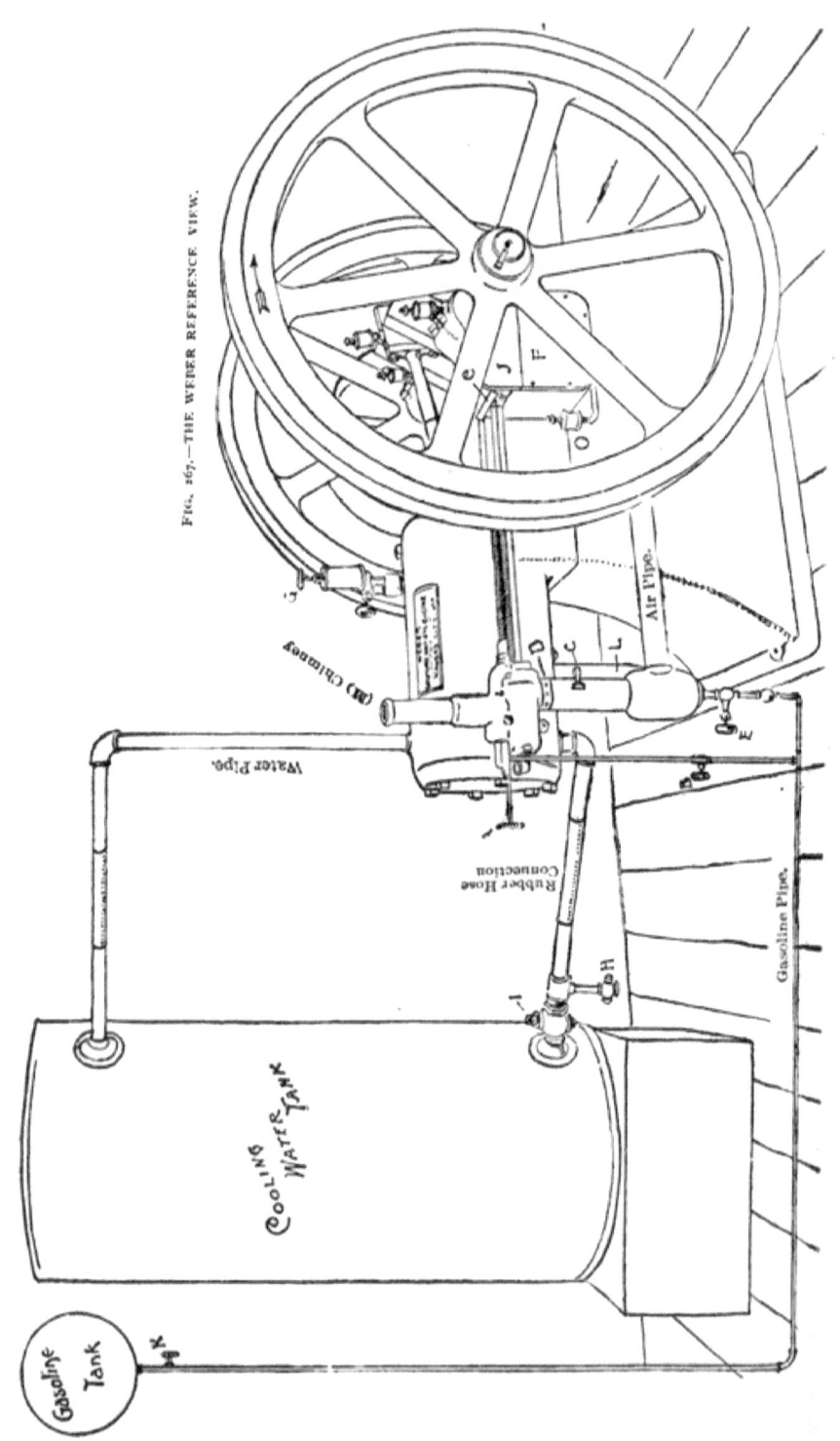

FIG. 167.—THE WEBER REFERENCE VIEW.

VARIOUS TYPES OF ENGINES AND MOTORS. 225

miners, quarrymen, and contractors. The engines of this company are also designed for the use of kerosene, crude oil, and distillate.

The style of horizontal engine (Fig. 166) of from 3 to 15

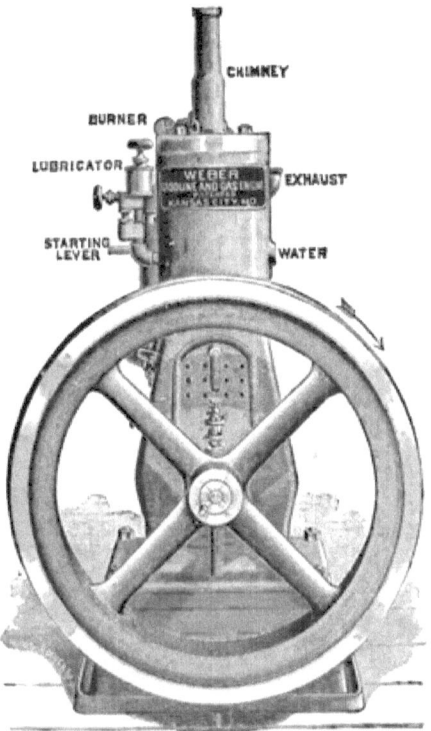

FIG. 168.—THE VERTICAL WEBER.

B.H.P. has three valve push-rods—the inner one opens the exhaust valve, the middle one opens the inlet valve, and the outside rod operates the timing-valve in the igniter passage.

Referring to the lettered diagram (Fig. 167), which is arranged for gasoline, A is the needle valve to the igniter burner, B the gasoline valve, C the handle of the gasoline mixing-valve, which is also the starting-lever for letting in the first

15

226 GAS, GASOLINE, AND OIL ENGINES.

FIG. 169.—THE WEBER GASOLINE HOISTING ENGINE.

VARIOUS TYPES OF ENGINES AND MOTORS. 227

FIG. 170.—THE WEBER GAS, GASOLINE AND CRUDE OIL ENGINE.

charge of gasoline. When the engine is running this valve is opened by the suction of the piston. In the larger engines it is counterweighted, as seen in Fig. 170. D is a collar for connecting the vaporizing pipe L; E, valve for regulating the gasoline supply; c, a lever to throw out the timing-valve when starting.

The governor on the smaller engines is of the pendulum type. It operates the inlet or charging valve, opening the valve at every other revolution at normal speed, and missing the contact at increased speed when the spring holds the valve closed until decreasing speed allows the governor to act on the push-rod and again open the inlet valve.

The governor on the larger engines is a fly-weight on the reducing-gear, adjusted by a spring and set nuts. O is a glass gauge to show the height of oil in the gear box; J is its cover.

In their latest style of engine (Fig. 170) the main exhaust is through ports in the cylinder opened by the piston at the termination of the stroke, with a supplementary exhaust valve in the cylinder head operated by a lever and push-rod. The timing-valve is operated by a lever pivoted on the cylinder, in contact with an adjustable push-block on the inlet-valve push-rod.

In the later designs of the Weber many improvements have been introduced to facilitate easy starting and for adapting it for pumping water for irrigation, for which purpose it is well suited and largely used. Its adaptation for the use of kerosene and heavy petroleum oils, and also for crude petroleum, has made it a very useful motive power for agricultural work.

The Priestman Oil Engine.

This has been long in use in Europe, and for several years past has been largely improved by the American builders, Priestman & Co., who have introduced a new system for perfecting the atomization of crude and kerosene oils, or any of the cheap distillates of petroleum. By the system adopted in

VARIOUS TYPES OF ENGINES AND MOTORS. 229

this engine, perfect combustion is produced; ignition is made positive, and the fouling of the cylinder and valves is obviated to such extent as to require cleaning only at periods of sev-

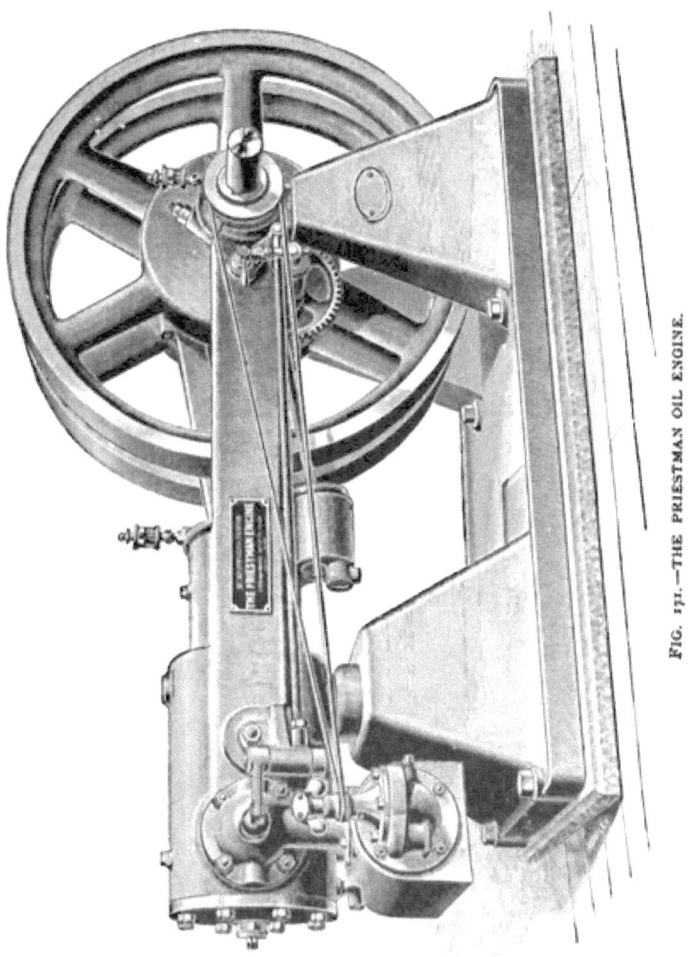

FIG. 171.—THE PRIESTMAN OIL ENGINE.

eral months. The low cost of the heavier petroleum distillates used makes the cost of power the lowest that can be obtained in an explosive motor.

In the cut, Fig. 172, A is the oil tank filled with any ordinary

230   GAS, GASOLINE, AND OIL ENGINES.

high test (usually 150° test) oil, from which oil under air pressure is forced through a pipe to the B three-way cock, and thence conveyed to the C atomizer, where the oil is met by a current of air

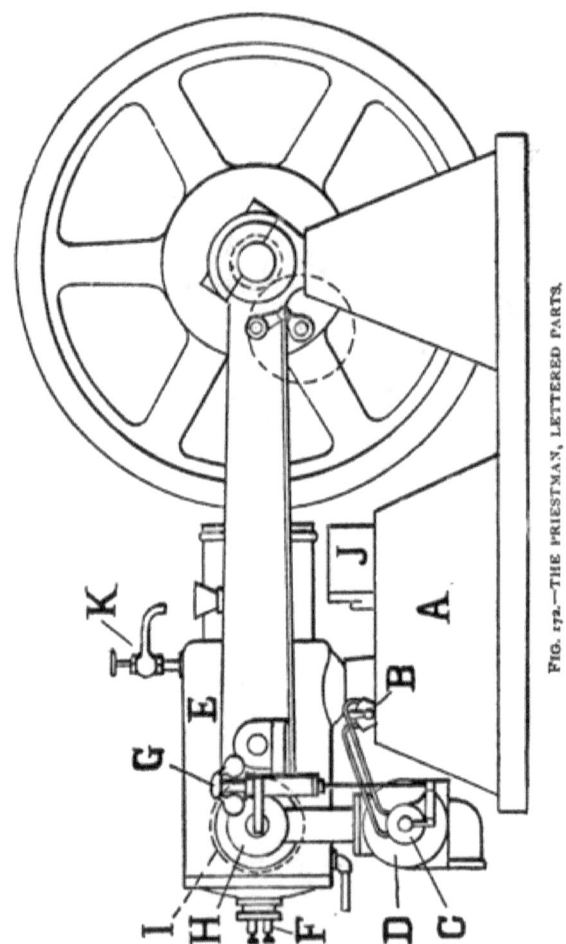

FIG. 172.—THE PRIESTMAN, LETTERED PARTS.

and broken up into atoms and sprayed into the D mixer, where it is mixed with the proper proportion of supplementary air and sufficiently heated by the exhaust from the cylinder passing around this chamber. The mixture is then drawn by suction through the I inlet valve into the E cylinder, where it is compressed by the piston and ignited by an electric spark passing between the points of the F ignition plug, the current for the spark being supplied from an ordinary battery furnished with the engine, the G governor controlling the supply of oil and air proportionately to the work performed. The burnt products are then discharged through the H exhaust valve, which is actuated by a cam. The I inlet valve is directly opposite the exhaust valve. The J air pump is used to maintain a small

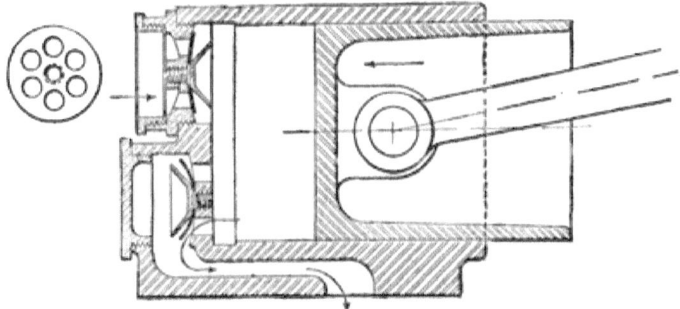

FIG. 173.—THE AIR PUMP.

pressure in the oil tank to form the spray. K is the water-jacket outlet.

Fig. 171 illustrates the general features of this engine. It is built on the straight-line principle, by which the moment of greatest strain from the power impulse is met by the frame in direct lines between the points of pressure.

The design is of the four-cycle compression type, with poppet valves, and its regulation is by varying or cutting off the supply of atomized oil. The oil fuel is placed in the base of the engine in an air-tight chamber, A in Fig. 172. A small air-pump, J, operated from the reducing-gear shaft forces air into the oil chamber with a pressure sufficient to cause the oil to be lifted to the three-way adjusting cock B, which also admits air from the compressed air in the oil tank; and oil and air pass to the atomizer through two small pipes, where their proportion and quantity are regulated by the governor.

The atomized oil and air are then injected into a jacketed cylinder, seen beneath the cylinder head and shown in section in Fig. 174, where it is completely vaporized by the heat from the exhaust in the outer chamber and further mixed with air to make a perfect explosive mixture by the indraught of air by the suction of the piston. The indraught of air by the suction of the piston is also regulated by the governor, and enters the

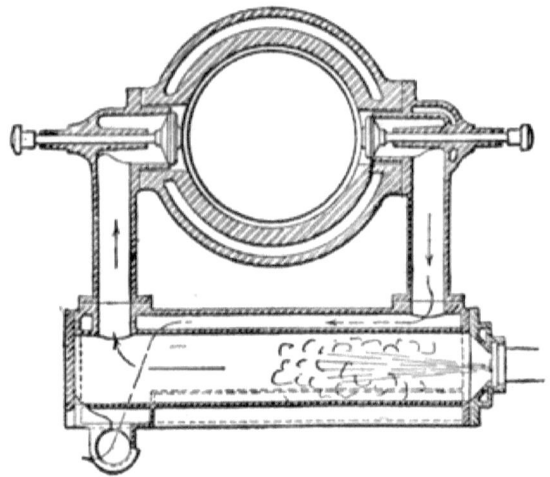

FIG. 174.—THE JACKET VAPORIZING CYLINDER, INLET AND EXHAUST VALVES.

vaporizing jacket cylinder in an annular stream around the atomized jet, as shown in Fig. 175, which represents a section of the governor and inlet passages. For starting the engine a small hand-pump is used for the first charge. The bottom of the inside chamber of the jacketed cylinder is heated to perfect the vaporization of the first charge by a lamp placed under the D-shaped cover seen in Fig. 171. In this engine the lubrication of the cylinder and piston is accomplished by the oil of the working charge. A new heat device has been lately introduced for ignition for the Priestman engines, which for some reasons is preferred to the electric igniter.

In Fig. 176 is represented an indicator card of the Priestman

VARIOUS TYPES OF ENGINES AND MOTORS. 233

engine, running under the three conditions of full load, half-load, and no load. The full line commences the compression

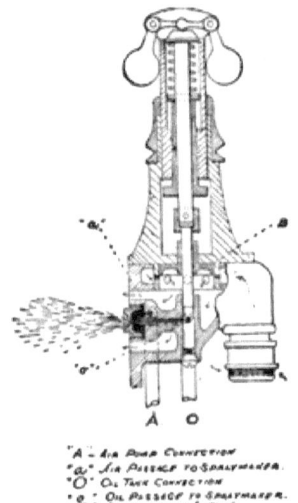

A - Air Pump Connection
a - Air Passage to Sprayholder
O - Oil Tank Connection
o - Oil Passage to Spraymaker
B - Supplementary Airtake

FIG. 175.—GOVERNOR AND ATOMIZER.

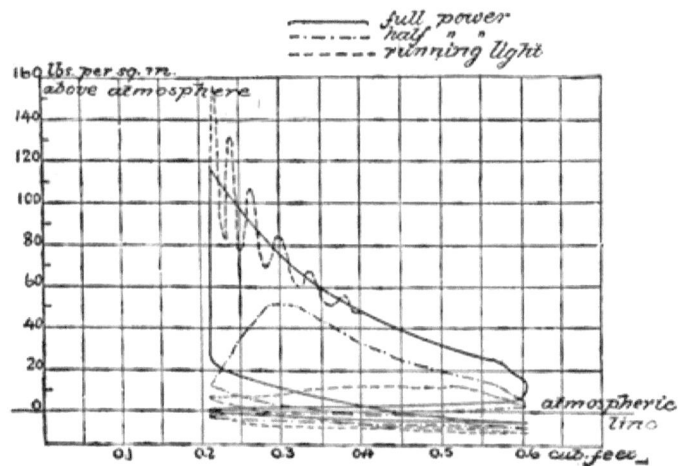

FIG. 176.—INDICATOR CARD OF THE PRIESTMAN OIL ENGINE.

at three-eighths of the stroke, and, with a clearance equal to one-half the piston stroke, the compression reaches 22 lbs. per square inch and is fired just before the termination of the compression stroke. The quick combustion is shown by the nearly vertical line, and its velocity is shown by the bound of the indicator arm above the mean, and its vibration continued, possibly helped by irregular combustion for one-half the stroke, as shown by the upper dotted lines, the continuous line showing the mean curve.

The second dotted line, showing a half-load card, indicates very clearly the retardation of combustion by weakening the charge of both oil and air, and the consequent lowering of all the lines of the card, carrying the charging line far below the atmospheric line. In the lowest and light-running card, the whole value of the card drops so as to make the card mean value about equal to the engine friction. It is certainly an interesting card for study, and we only wish that we could show this class of cards on a larger scale and for all the conditions of governing by limitation of fuel to compare with governing by closure of the exhaust valve.

The Lawson Gas and Gasoline Engine.

The Lawson engines are built by Welch & Lawson. They are of the four-cycle compression type and of the vertical style. They are built in eight sizes, from ½ to 15 B.H.P. with single cylinders, and of 20 and 30 B.H.P. with double cylinders. The concern also builds gasoline engines for horseless wagons and carriages. Figs. 177 and 178 represent two styles of the vertical engine. The valves have a positive motion from two sets of reducing-gear, Fig. 177, one of which operates the poppet-exhaust valve by a push-rod and cam on the reducing-gear shaft. The gas and air inlets are on the opposite side of the cylinder from the exhaust. The gas valve is a poppet, operated directly by a push-rod from a cam on the reducing-gear shaft, while a piston valve operated by a push-rod from a crank-pin on the reducing-gear governs the air inlet independently of the gas-inlet valve.

VARIOUS TYPES OF ENGINES AND MOTORS. 235

By this arrangement the air inlet is opened before the gas inlet is opened, and allows a sweep of pure air to enter at the head of the cylinder, followed by the mixture of gas and air; thus in a measure keeping the explosive mixture of gas and air

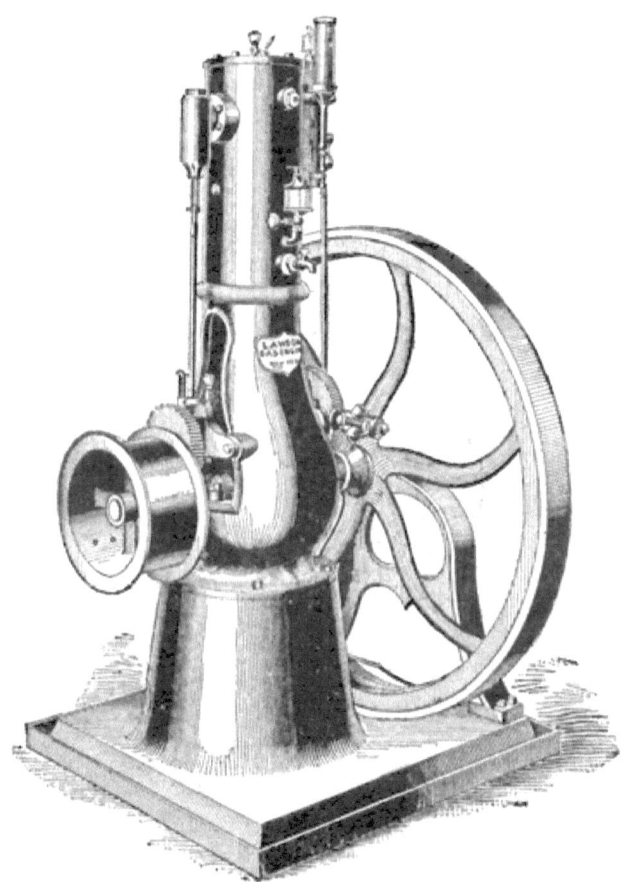

FIG. 177.—THE LAWSON VERTICAL.

separate from the products of the previous explosion by injecting it across and next to the cylinder head where the igniter inlet enters the cylinder. The same cycle of operation is made in the engine Fig. 178, by a single set of gearing.

The igniter is of the hot-tube style, entering the side of the cylinder directly under the head. The governor is of the horizontal, centrifugal style, taking its motion through a bevel gear

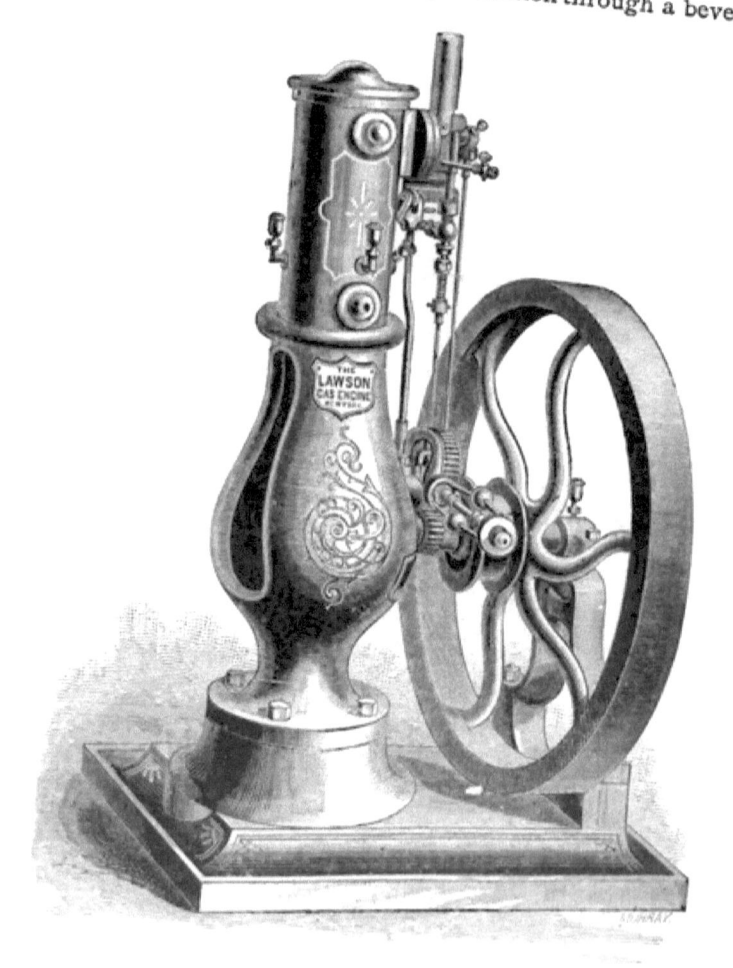

FIG. 178.—THE LAWSON AIR AND GAS VALVE GEARING.

from the reducing-gear shaft, and operates the gas-valve push-rod for a variable gas charge.

The Lawson pumping engines (Fig. 179) are made in two

sizes, 1 and 2 B.H.P. These engines are constructed on the same principles as the power engines, only with inverted cylinder and with pump attachments on a single square base.

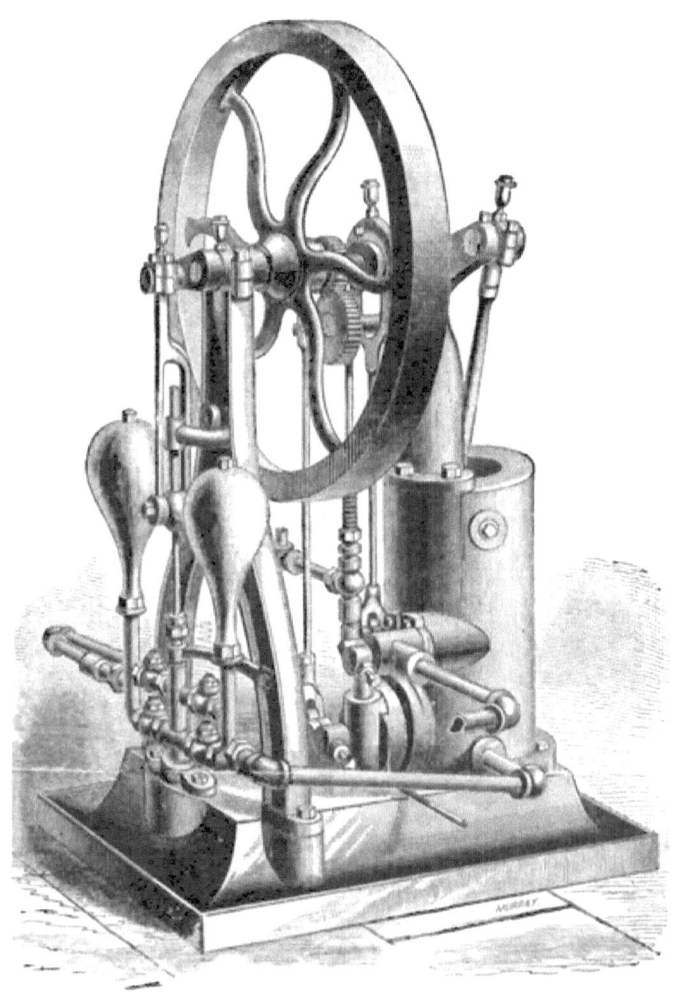

FIG. 179.—THE LAWSON PUMPING ENGINE.

This company is now building kerosene-oil engines of similar pattern as here described.

238   GAS, GASOLINE, AND OIL ENGINES.

The Racine Gas and Gasoline Engine.

The engines of the Racine Hardware Company combine some of the most recent improvements in construction. They

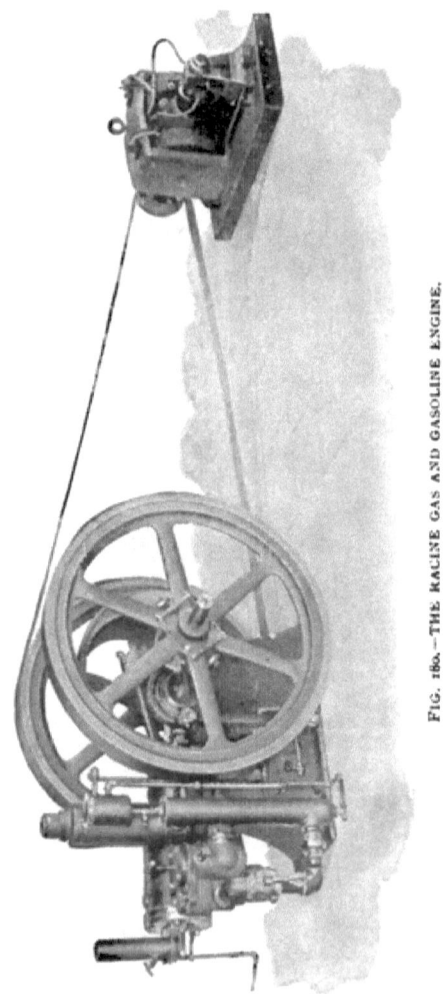

FIG. 180.—THE RACINE GAS AND GASOLINE ENGINE.

are of the four-cycle compression type. All valves are of the poppet style. The regulation of speed is made by a miss-opening of the exhaust valve, by which a fresh charge is excluded when the piston cushions on the previous charge until the normal speed is reached, when the governor again opens the exhaust valve and allows a fresh charge to be drawn in. This company furnishes both hot-tube and electric igniter for all their engines, so that failures shall not occur by the disabling of one or the other of the igniting apparatus.

The governor is of the horizontal centrifugal type, revolv-

FIG. 181.—THE RACINE GASOLINE ENGINE.

ing on the main shaft, and by a lever connection produces a lateral movement of a rolling disc attached to the lever of the exhaust push-rod. The lateral motion of the governor-controlled disc rides the disc on to or off the exhaust cam on the reducing-gear for a miss-exhaust. The gasoline pump is operated by a cam on a small shaft driven by the reducing-gear, and furnishes a surplus supply to a receiving cup over the mixing-chamber, with an overflow pipe returning the surplus gasoline to the tank by gravity. Between the supply cup and the mixing-chamber there is a sight-feed valve, by which the flow of gasoline to the mixing-chamber may be observed and regulated. Any surplus or overfeeding produces no dangerous conditions, as the gasoline entering the mixing-chamber in excess falls into the recess at the bottom and is conveyed back to the tank through the overflow pipe from the supply cup. It will be observed by inspection of the cuts (Figs. 181 and 182) that the exhaust pipe is jacketed for a short distance above the engine, with inlet holes for the entrance of air at the top and a neck from the jacket to the mixing-chamber below, so that the air is warmed before meeting the incoming gasoline in the mixing-chamber, where by an extended surface the gasoline is perfectly vaporized and mixed with air for best effect. The

FIG. 182.—THE RACINE GASOLINE ENGINE.

quantity drawn in for ignition is regulated by the index valve near the inlet valve, at which point a further admixture of air completes the proportions necessary for the desired explosive action.

At present these engines are built of 2, 3, and 4 B.H.P. They are well adapted for small electric-lighting plants, as shown in Fig. 180.

The Hornsby-Akroyd Oil Engine.

This engine is of English origin and now built by the sole licensees of the United States patents—the De La Vergne Refrigerating Machine Company—in all sizes from 4 to 55 H.P. They are of the four-cycle compression type, using any of the heavy mineral oils or kerosene as fuel.

This unique explosive engine seems to be a departure in design from all other forms of explosive engines, in the manner of producing vaporization of the heavy oils used for its fuel and the manner of ignition.

An extension of a chamber from the cylinder head, somewhat resembling a bottle with its neck next to the cylinder head, performs the function of both evaporator and exploder.

FIG. 183.—THE HORNSBY-AKROYD OIL ENGINE.

Otherwise these engines are built much on the same lines of design as gas and gasoline engines, having a screw reducing-gear and secondary shaft that drives the governor by bevel gear, the automatic cylinder lubricator by belt, and cams for operating the exhaust valve and oil pump.

The bottle-shaped extension is covered in by a hood to facilitate its heating by a lamp or air-blowpipe, and so arranged as to be entirely closed after the engine is started, when the red heat of the bottle or retort is kept up by the heat of combustion within. The narrow neck between the bottle and cylinder, by its exact adjustment of size and length, perfectly controls the time of ignition, so that of many indicator-cards inspected by the writer there is no perceptible variation in the time of ignition, giving as they do a sharp corner at the compression terminal, a quick and nearly vertical line of combustion, and an expansion curve above the adiabatic, equivalent to an extra high mean engine pressure for explosive engines.

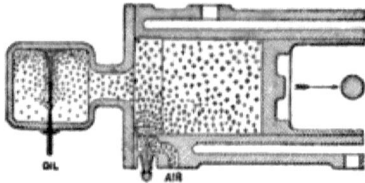

FIG. 184.—INJECTION, AIR AND OIL.

The oil is injected into the retort in liquid form by the action of the pump at the proper time to meet the impulse stroke,

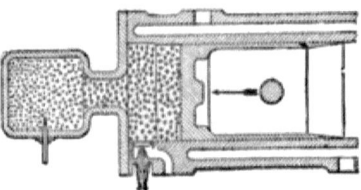

FIG. 185.—COMPRESSION.

and in quantity regulated by the governor. During the outer stroke of the piston air is drawn into the cylinder and the oil is

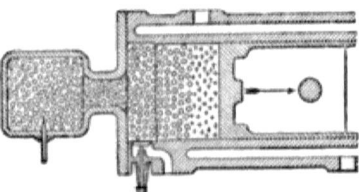

FIG. 186.—COMBUSTION AND EXPANSION.

vaporized in the hot retort. At the end of the charging stroke there is oil vapor in the retort and pure air in the cylinder, but non-explosive. On the compression stroke of the piston the air is forced from the cylinder through the communicating

VARIOUS TYPES OF ENGINES AND MOTORS. 243

neck into the retort, giving the conditions represented in Fig. 184 and Fig. 185, in which the small stars denote the fresh air entering, and the small circles the vaporized oil. In Fig. 185 mixture commences, and in Fig. 186 combustion has taken place, and during expansion the supposed condition is repre-

FIG. 187.—THE HORNSBY-AKROYD PORTABLE ENGINE.

sented by the small squares. At the return stroke the whole volume of the cylinder is swept out at the exhaust, and the pressure in the retort neutralized and ready for another charge.

It is noticed by this operation that ignition takes place within the retort, the piston being protected by a layer of pure air.

It is not claimed that these diagrams are exact representations of what actually takes place within the cylinder; nevertheless, their substantial correctness seems to be indicated by the fact that the piston rings do not become clogged with tarry substances, as might be expected.

This has been accounted for by an analysis of the products of combustion, which shows an excess of oxygen as unburned air; which indicates that the oil vapor is completely burned in the cylinder, with excess of oxygen.

In Fig. 187 is illustrated the adaptation of this engine for portable power. It is largely in use for electric work, for air compressing, ice machinery, and pumping. The United States Light-House Department has adopted this engine for compressing air for fog whistles. Traction engines and oil-engine locomotives for narrow-gauge tramways and mining railways will soon be one of the manufacturing departments of the De La Vergne Company.

The Climax Gas Engine,

made by the Climax Gas Engine Company, is of the four-cycle compression type, with globular combustion chamber. The

FIG. 188.—THE CLIMAX GAS ENGINE.

VARIOUS TYPES OF ENGINES AND MOTORS. 245

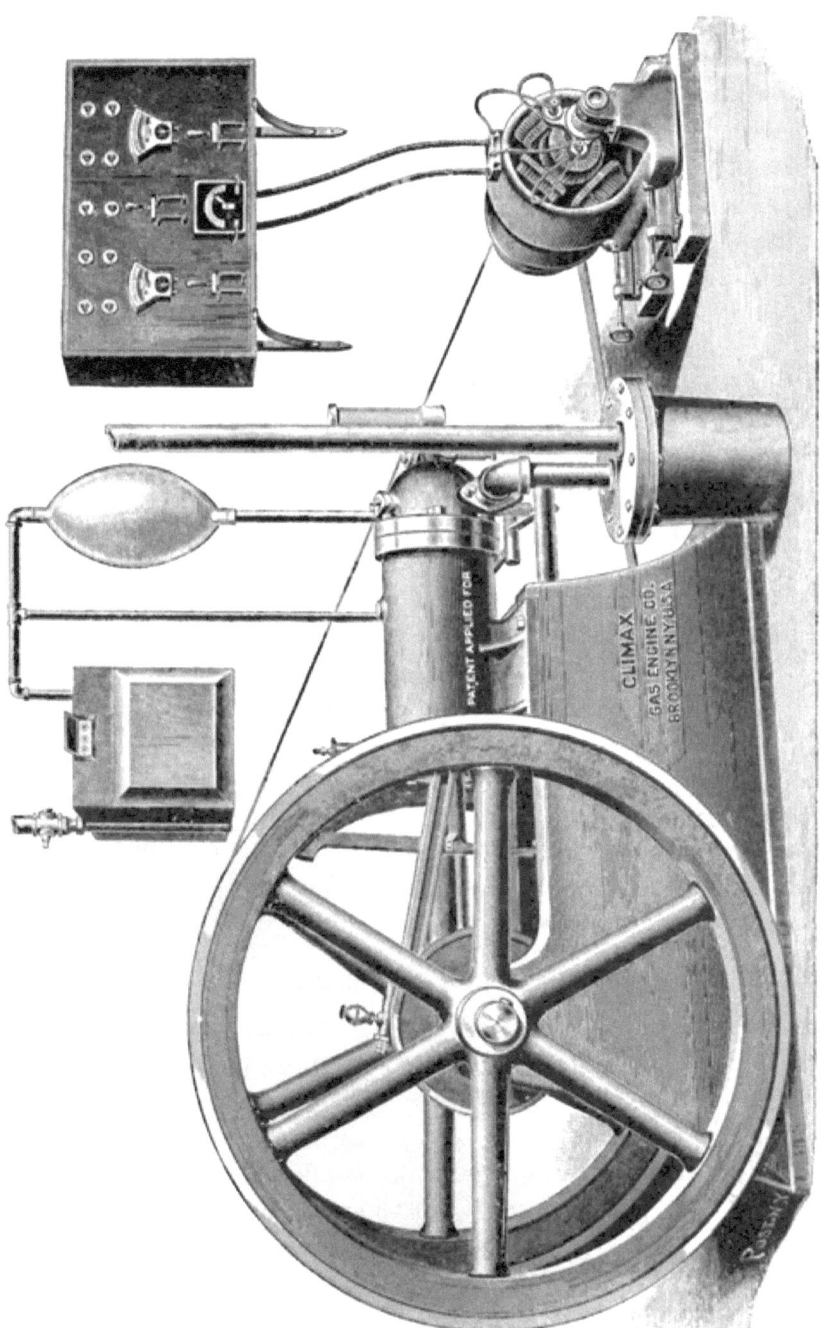

FIG. 189.—THE CLIMAX GAS ENGINE AND ELECTRIC LIGHT PLANT.

air and gas inlet is at the end of the globular cylinder head, to which is inserted and attached all the valves and valve gear. The valve-gear shaft is driven by a worm gear from the engine shaft, and carries a cam for operating the exhaust valve through a lever. A cam at the end of the cam-shaft operates an inertia governor, which by its momentum makes a hit-or-miss opening of the gas-inlet valve as required by the speed of the engine. The governor is made adjustable while the engine is running by turning a milled-head screw and tightening or relieving the tension of a spiral spring that controls the momentum of the governor bob.

The regulation of the gas flow is made by an index valve close to the inlet valve. The globular cylinder head has a water circulation. Hot-tube ignition, with automatic self-starting attachment, are on the larger size engines. The engines of this company are made in nine sizes for stock, from 1¼ to 40 B.H.P. Engines of any desired horse-power larger than 40 B.H.P. are made to order.

These engines are well adapted for electric lighting, and the Climax Company guarantees the electrical output on the measured gas consumption.

In electrical light trials with this engine, the variation by the sudden shutting off of a quarter, half, or three-quarters of the number of lamps shows an oscillation of less than two volts, and with a gas consumption not exceeding 40 cubic feet per kilowatt per hour.

The New York Motor.

This is one of the new style high-speed motors of light weight, weighing but 150 lbs. for a 1½ H.P. motor, including the fly-wheel. It is made by the New York Motor Company. It is operated by gas, gasoline, or carbonated oil. The stationary style, as shown in Fig. 190, has the water tank directly over the engine on a frame, which also holds the battery and sparking-coil. By the direct and close water connection the

VARIOUS TYPES OF ENGINES AND MOTORS. 247

water in the tank becomes warm, and by its rapid circulation keeps the cylinder at the proper temperature for economic consumption of gas or other fuel—the slow evaporation from the

FIG. 190.—THE NEW YORK MOTOR.

open top of the tank being sufficient to keep the water at an even temperature of about 180° F.

Several novel features are claimed in its construction. The crank is encased and runs in an oil bath, thus keeping crank and piston lubricated. The shaft has an outboard bearing, which counteracts the belt strain. The motion of the piston is made to produce an air circulation in the piston and lower part of the cylinder to prevent undue heating, thus keeping the piston and cylinder at a uniform temperature.

The inlet valve is so constructed that the new charge is conducted directly down to the piston, and on compression the spark flashes in the centre of the combustion chamber, making

FIG. 191.—THE NEW YORK MOTOR.

a quicker explosion and keeping the electrodes free from fouling.

The valve mechanism is very simple and of the poppet kind, consisting of one double valve, operated by one cam, one roller, and one slide. Both valve and igniter are operated by cams on a reducing-gear wheel. Both electric and hot-tube igniters are used, as preferred.

The gas and air charges are regulated by index valves, with an additional control of the gas charge by a ball governor running by belt from the main shaft. For a launch a friction-clutch for reversing the propeller wheel is used. This is one of the few very light-weight and high-speed engines adapted for small power and portability.

The Facile Oil Engine.

Originally built by the Britannia Company, Colchester, England, and now built in the United States by Mr. John A.

FIG. 142.—THE FACILE OIL ENGINE.

Holmes, who controls the United States patents and is bringing out the general features of the English engine with modification and improvements derived from experience and the needs of a perfect motor, using the heavy oils and kerosene as explosive fuel.

In Fig. 193 we illustrate the vertical style as used for marine and vehicle propulsion. It is of the two-cycle compres-

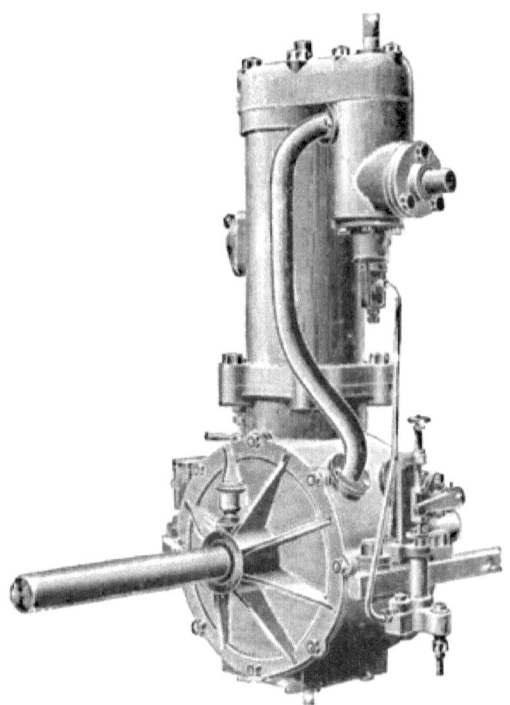

FIG. 193.—THE VERTICAL FACILE MARINE ENGINE.

sion type, and has but one valve, which by its peculiar construction operates as both inlet and exhaust valve. The valve is a ported piston, capped by a disc valve to hold the ports in their proper position and close the exhaust during the pressure stroke.

The crank chamber is closed, and by the downward stroke of the piston produces an air pressure that charges the combustion chamber at every revolution. It is self-igniting. The small pump seen in front, driven by a cam on the main shaft through a rock shaft and arms, with an adjusting screw to regulate the stroke, sends the oil into a small chamber seen in the extension below the combustion chamber, where it is vaporized by first heating the small chamber with a lamp to start with, after which the heat is retained by a tube extending up into the combustion chamber, when the lamp is removed and the operation of the engine becomes continuous automatically.

In Fig. 192 is illustrated a horizontal Facile engine, in which the two-cycle impulse is obtained by a differential action of the piston from its reduced size at the crank end operating through a stuffing-box, as seen in the cut. This engine has a separate valve chamber for the exhaust and inlet, which is controlled by a single valve, a combination of a ported piston and seated disc. Its operation is regulated by a secondary shaft and vertical centrifugal governor, which varies the charge.

These engines are built at present in a number of sizes, from 1 to 25 H.P., single and double cylinder.

The Simplex Naphtha Launch Engine.

A new engine, designed especially for boat service, has just been put on the market by Charles P. Willard & Co. These engines are of the two-cycle compression type, or with an impulse at each revolution of the crank. It is very simple in construction, receives its charge and exhausts through cylinder ports opened and closed by the movement of the piston at the end of the downward stroke.

A single eccentric on the main shaft operates, through a lever and two cams, the electric igniter alternately for forward and backward motion of the engine.

The valve seen on the cylinder regulates the charge from the closed-crank chamber, which is compressed by the downward stroke of the piston. The naphtha vapor and air are drawn into the crank case by the upward stroke of the piston, thoroughly mixed by the motion of the crank, and receives its maximum compression at the moment of opening the inlet port, when the compressed mixture rushes into the combustion chamber of the cylinder, while the exhaust port is still open to clear the cylinder of the products of the previous explosion.

FIG. 194.—THE SIMPLEX BOAT ENGINE.

These engines are built in sizes of 2, 4, and 6 H.P. The 2 H.P. engine weighs 300 lbs., and is suitable for a boat from 16 to 22 feet long. The 4 H.P. engine is suited for a boat 20 to 28 feet long, and weighs 500 lbs. All the engines run at a speed suitable for boat service up to 300 revolutions per minute.

The White & Middleton Gas Engine.

This engine is equally suited to both gas and gasoline, and is made by the White & Middleton Gas Engine Company. All their engines are of the four-cycle compression type, with the principal exhaust ports opened by the piston at the end of its

FIG. 195.—THE WHITE & MIDDLETON ENGINE.

explosive stroke, and with an additional or clearance-exhaust valve in the cylinder head.

The valves are all of the poppet type. The supplementary exhaust valve is operated by a lever across the cylinder head and a push-rod direct from a differential slide mechanism, which does away with the reducing-gear used on other engines. An arm on the push-rod operates the gas-valve stem, which is provided with a regulating adjustment.

The small roller disc on the push-rod mechanism is under the control of a centrifugal governor and a spring, being thrown out of gear with the shaft cam whenever the speed of the engine exceeds the normal rate, and thus failing to open the gas supply and the supplementary exhaust valve until the speed of the engine has returned to its normal rate. There is a relief valve opening into the supplementary exhaust passage for relieving the pressure in the cylinder when starting the

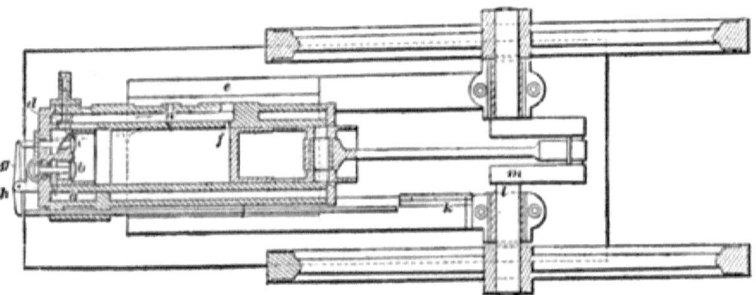

FIG. 196.—SECTIONAL PLAN OF THE WHITE & MIDDLETON ENGINE.

engine. The whole design of the engine is exceedingly simple and its action noiseless.

When gasoline is used the gas-supply valve is replaced by a small pump, which is operated by the push-rod, and its hit-or-miss stroke is governed by the action of the push-rod and its governor.

These engines are built in nine sizes, from 4 to 50 B.H.P.

The Hydrocarbon Motor and Launch.

The Hydrocarbon Launch Company are builders of open launches, cabin cruisers, and yacht tenders, equipped with an approved pattern of kerosene motors—which for cruising is claimed to be the most desirable for fuel, as kerosene is not only cheap, but can be purchased in every grocery store on the line of a cruise.

The boats are of fine lines and high finish for comfort and convenience, and of sizes of 16, 18, 21, 25, 30, 36, 42, 45, and

VARIOUS TYPES OF ENGINES AND MOTORS. 255

FIG. 197.—THE HYDROCARBON LAUNCH CO.'S, 18-FOOT LAUNCH—1 H.P. MOTOR.

256 GAS, GASOLINE, AND OIL ENGINES.

FIG. 198.—THE HYDROCARBON MOTOR AND REVERSING WHEEL.

VARIOUS TYPES OF ENGINES AND MOTORS. 257

FIG. 199.—POSITION OF MOTOR IN LAUNCH.

17

50 feet in length, with motors of suitable power for any desired speed.

The management of the motor is all done by direct connection at the wheel or tiller. Ignition is electric, and the motor can be started by charging the cylinder without turning the fly-wheel. Speed regulation is made by varying the charge in quantity but not in quality, so that explosions are obtained at every revolution of the wheel, whether running at full power or light. The kerosene fuel is injected into the combustion chamber in exact quantities for each explosion by a small pump, the stroke of which is handled by the steersman.

The new motor of this company is somewhat different from the one shown in our illustration. It has been reduced to the simplest terms in its working parts, to better adapt it for use by persons not posted in the details of motor engineering. The motor and propeller run constantly in one direction, and the various movements of the propeller blades for forward, slow, stop, and backing are controlled by a lever at the tiller or steering-wheel.

The following table gives the sizes of launches, motors, capacities, and cost of running as made by the Hydrocarbon Launch Company:

Length.	Motor.	Draught.	Beam.		Depth.		Passengers carried.	Speed per hour.	Cost per hour.
Feet.	H. P.	Inch.	Ft.	In.	Ft.	In.		Miles.	Cents.
16	$\frac{3}{4}$	12	4			18	4 to 6	5 to 5$\frac{1}{2}$	$\frac{3}{4}$
16	1	16	4	8	2		5 to 7	5 to 6	1
18	1	18	5		2	2	6 to 10	6 to 7	1
21	2	22	5	6	2	3	10 to 15	6$\frac{1}{2}$ to 7$\frac{1}{2}$	2
25	4	24	6		2	6	15 to 20	7$\frac{1}{2}$ to 8$\frac{1}{2}$	4
30	7	27	6	6	2	10	20 to 25	8$\frac{1}{2}$ to 9$\frac{1}{2}$	7
33	7	28	7		3	2	22 to 28	9 to 10	7
35	12	30	8		3	6	25 to 30	10 to 11	12
40	12	34	8	6	3	8	30 to 35	10 to 12	12
50	Two 12	38	9	6	4	2	35 to 45	11 to 14	24

The Duryea Motor Wagon.

Fig. 200 illustrates the general appearance of the motor wagon made by the Duryea Motor Wagon Company. Their motor wagons were the winners of prizes in the Chicago races of 1895 and in the Cosmopolitan race of 1896. It has 34-inch front and 38-inch rear wheels, with $2\frac{1}{2}$-inch pneumatic tires, is steady in action, and easy and comfortable to ride in. Its low rig makes it a most desirable vehicle for a physician or for messenger service, a most convenient carriage for ladies for

FIG. 200.—THE DURYEA MOTOR WAGON.

park or road riding. It runs backward or forward with equal facility—backward at 3-mile speed, and forward at 5-, 10-, and 20-mile speed. It has two independent motors of about 3 H.P. each, so that with any derangement of one motor the other is available for ordinary speed. It uses electric exploders. It is speeded and guided by the vertical and horizontal motion of a single lever; carries 8 gallons of gasoline, sufficient for a trip of 100 or 200 miles.

The steering action is so arranged that obstructions will not jerk the lever from the hand.

The Gasoline Motor Bicycle.

In Figs. 201 to 204 is illustrated a German gasoline motor bicycle, made by Wolfmuller & Geisenhof, Munich, Germany. A large number of bicycles of this type are in use in Munich and Paris. It is similar in type to the lady's bicycle, being easy to mount and start without mishaps, from its low centre of gravity. The hind wheel is composed of two sheets of thin steel and a rim, which gives it great stability under the load, the machine alone weighing 110 lbs. It is actuated by two pistons, and is equal to 2 H.P. The speed can

FIG. 201.—THE MOTOR BICYCLE.

be regulated from 3 to 24 miles per hour. All the operations for controlling speed, guiding, and the brake are constantly in the hands. The gasoline tank is placed between the tube frames, and contains gasoline sufficient for a trip of 100 miles.

All the essential parts are placed in the interior of the frame, and are consequently protected against damages caused by a collision, fall, etc.

The gasoline reservoir M is located behind the head of the bicycle, and may be filled through the tubulure *m*, with a quantity of liquid sufficient for 120 miles. The gasoline falls drop by drop into the evaporator N, in passing through the cock S and the funnel T. Through a simple mechanism, shown in Fig. 204 (4), the gas mixed with air in proper proportions enters the ignition chamber through the valves O. (2)

A lamp P, which continually keeps at a red heat a small tube p, placed above the flame, produces the explosion of the detonating mixture. The piston I is thus driven into the cyl-

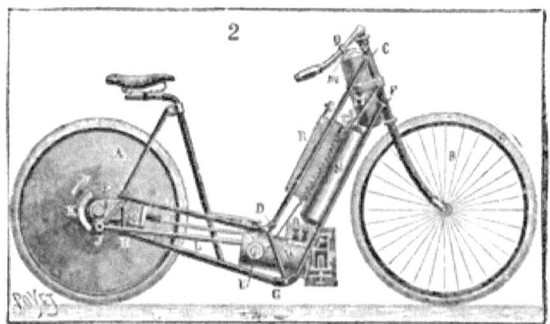

FIG. 202.—DETAILS OF THE MOTOR BICYCLE (ELEVATION).

inder W, and actuates around the axis I the rod I J, which is aided in its return motion by a powerful spring, E J.

The most important control given to the handle-bar piece

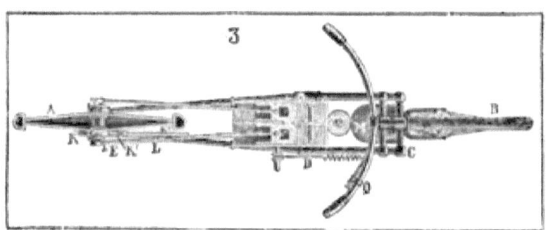

FIG. 203.—DETAIL PLAN.

Details of the Motor Bicycle (Figs. 202 and 203). A, Driving wheel; B, steering wheel; C, D, E, F, G, H, frame tubes; M, gasoline reservoir; N, evaporator; O, valve box; P, lamp and ignition chamber; p, ignition tube; R, water reservoir; S, cock for regulating the entrance of gasoline into the evaporator; T, funnel of the evaporator; U, regulator of water for cooling the cylinders; V, distributing mechanism; W, cylinders; I J, connecting rod; K, cam; K', roller; K", rod of the distributing mechanism; L, piston.

is the entrance and exit of the evaporator N. The latter is thus named because the gasoline, falling drop by drop through the funnel T, evaporates therein. A succession of gauze sieves $a\ a'$, etc., placed one above another in the cylinder, offers therein the greatest surface of evaporation possible. The external air, which through its mixture with the gas is to produce the detonating mixture, enters the cylinder through b and the pipe b', through a capsule that prevents the suction of impurities and dust. The admission of the mixture into the valve chamber is regulated by the piston c, whose rod d is placed, like the

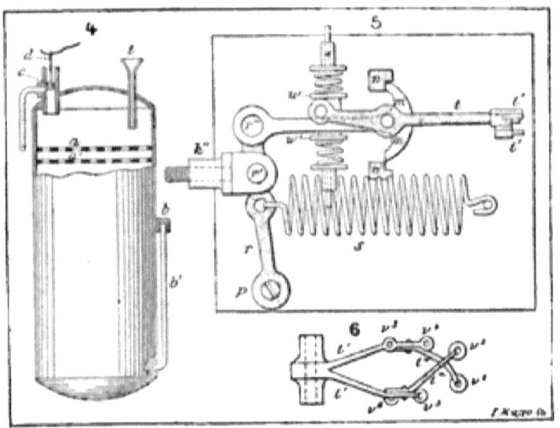

FIG. 204.—DETAILS OF THE EVAPORATOR, ETC.

Details of the Evaporator—partial section: t, Funnel for entrance of gas; a, a', etc., gauze for accelerating evaporation; b, b', tubes for entrance of the air; c, piston for admitting the mixture into the valve box; d, its rod. 5, Details of the distributing Mechanism: K', Extremity of the actuating rod; r, t, levers; p, r', r', joints; s, spiral spring; w, w', supports of the spring; n, n', stop-blocks. 6, Details of the various Valves: v^1, v^2, Ignition valves; v^3, suction valve; v^4, v^5, emission valves; v^6, air valve.

gasoline cock, under the absolute control of the rider. If, then, the latter completely closes the cock, he thus also hermetically closes the admission tube at the same time. The gasoline ceases to fall upon the gauzes and the mixture to enter the ignition chamber, and conversely. The cam K, fixed upon the disc wheel A and carried along in its revolution, frees, in passing, the roller K', mounted upon a guide block that transmits motion to the traction rod K". It is this rod that, at V, actuates the distributing mechanism, which it is impossible to represent in Figs. 202, 203; the principal details of which are shown in Fig. 204 (5) (6). This mechanism is installed upon a plate that forms a cover for the cooling-box of the cylinders. It is constructed as follows: The extremity of the rod K' is jointed at r' with a lever r, that oscillates around the fixed point p, and is continuously brought back to its normal position by a powerful spring S as soon as the passage of the cam K over the roller K' has made it lose it.

The extremity of this lever r is jointed at r'' to another lever t, whose extremity commands, at t', the valves represented in Fig. 202. At about its centre the lever t is jointed again to a crosshead m, and held upon it with hard friction by two spiral springs. This head engages with the blocks n and n', which are provided with corresponding notches. The central part of the lever t is thrust alternately against w and w'. On another hand, the levers, t' (Fig. 204) (6) carry at their extremity another small lever, t'' which controls the valves v^a and v^a, leading to the ignition chamber. Owing to this arrangement, the lever t' of one of the cylinders causes at the same time the ignition in the conjoined cylinder.

If now we suppose that the cam K carries along the rod K', it will be seen that the lever t will recoil and carry with it one of the levers, t'. The crosshead m engages at the same time with the block n, and compresses the spiral spring which is located behind the piece, w. But as soon as the powerful spring S acts, it brings the lever t to the front and causes the head m to engage at n, carrying with it the second lever $t\ t'$, and reciprocally.

It is certain that the complication of the pieces is here very formidable for a machine designed for a little of every kind of speed and all kinds of roads, but we must also remember that we are as yet witnessing only the first trial of automobile cycling, and we ought to give the inventors a margin of some time. However it be with the criticisms of detail that we might formulate, one fact remains, and that is that the bicycle that we have described is really in operation. Its success in Germany and Switzerland is already so great that the entire product of the manufacturers has been engaged. A number of these motor bicycles are now in use in the United States.

The Bollee Automobile Tricycle.

The Bollee tricycle is a French gasoline carriage of the bicycle type, built by Mr. Leo Bollee, of Mans, France. The engine is of the four-cycle type, single cylinder, of more than

FIG. 205.—THE BOLLEE TRICYCLE.

usual length, designed to carry the expansion as far as practicable. Gasoline vapor is produced in a carburetter.

The engine is of 2 H.P., and makes about 800 crank revolutions per minute at full speed (27 miles per hour), and operates the vehicle axle by belts and friction-clutches, producing a noiseless motion of the machinery, with the attenuated exhaust smothered by a muffler. The whole apparatus, weighing only 350 lbs., is most conveniently arranged for quickly mounting, and with all the driving and steering gear under the immediate control of one hand.

The slight elevation of the vehicle gives it a perfect stability, since its centre of gravity is situated but 16 inches above the surface of the ground. Its wheel base is 3½ by 4 feet.

The steersman sits behind, his feet resting on each side upon a platform provided with a straw mat. He merely has to move his foot backward in order to press the lever of a powerful brake, whose block is tangent to the circumference of the driving-wheel. With his right hand he steers the vehicle through a hand wheel, which, by a very simple gearing, turns the fore wheels to the right or left. With the left hand he holds an almost vertical lever, which permits him with a few motions to effect several important manœuvres. If he pushes it forward he tautens the driving-belt, and consequently starts the vehicle as soon as the motor has been set in operation through a winch, according to the well-known process. If, in the median position of the lever, he turns the handle to the right or left, he throws the motor into gear into one or another of the three speeds. Finally, if he pulls the lever backward, he loosens the belt and consequently suppresses the transmission, and, at the same time, presses the brake block against the driving-wheel.

Wing's Gas and Gasoline Engine.

The Wing engines are especially designed for marine and vehicle propulsion, being very light and compact.

They are of the four-cycle compression type, are made single and duplex cylinder for marine propulsion, and tandem cylinder for vehicle service, the cylinders for vehicles being placed end to end with the shaft between. Cylinders are of 4-inch diameter, with 5-inch stroke.

Fig. 206 illustrates the general features of this compact engine. The piston rod is guided by the bearing and stuffing-box in the piston head, which by enclosing the arm of the upper part of the cylinder makes it an air-compressor for blowing a whistle or for any other purpose for which compressed air may be needed.

The reducing-gear at the rear in the cut, and not shown, operates, by cams on a cross shaft, both the inlet and exhaust valve, making the valve action positive.

The vaporizer shown opposite the base at the left is a

FIG. 206.—WING'S MARINE GASOLINE ENGINE.

chamber with an inlet nozzle for the gasoline and a needle valve operated by the push-rod of the inlet valve. A check valve on both gasoline and air pipe prevents back-firing.

The main shaft has double cranks and the connecting rods are very light, there being two to divide the pulling strain. The general arrangement of the parts brings the weight of the engine very low in a boat—a most desirable feature.

Relief valves are provided for both chambers of the cylinder, so that starting is very easy.

Ignition is by electric spark caused by a pair of wiping-electrodes, the revolving one being operated by a small reducing-gear seen on the front of the cut. The small driving gear is revolved by a slotted arm and a pin fixed in the connecting rod.

The spring that returns the electrode to its position after contact is on the outside, and is not subject to heat.

The small pump in front at the lower side of the wheel is the water-circulating pump.

The boat engine in the cut is of 2 H.P., nominal, weighs 125 lbs., and is suitable for a 21-foot boat with a 16-inch wheel. It will run 500 revolutions per minute, and develop $2\frac{1}{2}$ B.H.P., with a boat speed of between eight and nine miles per hour.

These engines are built at present of 2 and 6 H.P. single cylinder and of 4 and 12 H.P. double cylinder.

PATENTS

Issued in the United States for Gas, Gasoline and Oil Engines and their appliances, from 1875 to 1896 inclusive:

— 1875 —
G. W. Daimler............... 168,623
J. Taggart.................... 161,454
P. Vera....................... 160,130

— 1876 —
J. Brady...................... 176,588
A. de Bischop................ 178,121
T. W. Gilles.................. 179,782

— 1877 —
J. Wortheim.................. 192,206
R. D. Bradley................ 187,092
F. Deickman.................. 195,585
N. A. Otto.................... 194,047
Otto & Crossley.............. 196,473

— 1878 —
J. Brady...................... 200,970

— 1879 —
F. Burger..................... { 222,569
                               { 222,660
J. H. Connelly............... 211,836
J. Robson..................... 220,174
Wittig & Hees................ 213,539
G. W. Daimler................ 222,467

— 1880 —
E. Buss....................... 226,972
L. Durand.................... 232,808
C. Linford................... 232,987
A. K. Rider................... 233,804
Wittig & Hees................ 225,778
D. Clerk...................... 230,470
G. W. Daimler................ 232,243

— 1881 —
E. Renier..................... 247,741
C. J. B. Gaume............... 240,994
A. K. Rider................... 245,218
J. Robson..................... 243,795
G. Wacker.................... 242,401

N. A. Otto.................... 241,707
J. Ravel...................... 236,258

— 1882 —
C. G. Beechy................. 264,126
R. Hutchinson............... 253,709
A. P. Massey................. 260,587
T. McAdoo.................... 253,406
P. Munsinger................. 266,304
L. C. Parker.................. 269,813
C. M. Sombart................ 260,620
K. Teichman................. 269,163
H. Wiedling.................. { 259,736
                              { 269,146
A. K. Rider................... 267,458
E. W. Kellogg................ 265,423
H. H. Burritt................ 258,884
W. H. Wigmore............... 260,513

— 1883 —
C. W. Baldwin............... { 276,747
                              { 276,748
                              { 276,749
                              { 276,750
                              { 276,751
                              { 287,897
                              { 288,399
                              { 290,310
J. Charter.................... { 270,202
                               { 270,203
H. Denney.................... 290,632
Eteve & Lallemont............ 272,130
J. A. Ewins................... 278,421
E. J. Frost................... 273,269
W. Hammerschmidt........... 288,632
Geo. M. Hopkins............. { 284,555
                              { 284,556
                              { 284,557
G. M. & L. N. Hopkins....... 284,851
Jackson & Kirkpatrick....... 283,398
S. Marcus.................... 286,030
H. S. Maxim.................. { 273,750
                               { 279,657

L. N. Nash.................. 271,902
                             278,255
                             278,256
                             289,019
                             289,691
                             289,692
                             289,693
N. A. Otto.................. 288,479
L. C. Parker (reissue)...... 10,290
G. H. Reynolds.............. 284,061
                             284,328
                             287,578
J. Robson................... 278,600
C. Rohn..................... 280,083
C. Shelburne................ 277,618
T. W. Turner................ 289,362
L. C. Parker................ 287,855

— 1884 —

G. M. Allen................. 301,320
J. Atkinson................. 306,712
J. Charter.................. 292,894
E. Edwards.................. 300,453
C. J. B. Gaume.............. 302,478
Geo. M. Hopkins............. 306,254
G. M. & I. N. Hopkins....... 305,452
I. N. Hopkins............... 306,924
C. W. King & A. W. Cliff.... 293,179
S. Lawson................... 306,933
                             307,057
H. S. Maxim................. 295,784
                             296,340
J. A. Menck—A. Hambrock..... 295,415
P. Murray, Jr............... 305,464
                             305,465
                             305,466
                             305,467
B. Parker................... 308,572
F. W. Rachholds............. 301,009
J. Spiel.................... 302,045
W. L. Tobey................. 306,443
S. L. Wiegand............... 297,329
J. S. Wood.................. 300,294
A. K. Rider................. 292,178
C. G. Beechey............... 306,314
S. Marcus................... 306,339
H. S. Maxim................. 296,341
                             291,065
                             293,762
                             302,271
                             293,185
J. Spiel.................... 291,102
C. H. Andrews............... 301,078

J. Schweizer................ 292,864
N. H. Thompson & C. B. Swan. 300,661

— 1885 —

S. Wilcox................... 332,312
C. H. Andrews............... 314,284
C. W. Baldwin............... 325,377
                             325,378
                             325,379
                             325,380
C. Benz..................... 316,868
M. G. Crane................. 327,866
G. Daimler.................. 313,922
                             313,923
W. A. Graham................ 330,317
H. Hârtig................... 324,554
G. M. & I. N. Hopkins....... 326,561
                             326,562
T. McDonough................ 315,808
L. N. Nash.................. 312,494
                             312,496
                             312,497
                             312,498
J. F. Place................. 322,477
                             328,970
D. S. Regan................. 333,336
C. Shelburne................ 322,650
                             332,447
D. S. Troy.................. 317,892
S. Wilcox................... 332,313
                             332,314
                             332,315
J. S. Wood.................. 328,170
A. W. Schleicher............ 314,727
H. P. Feister............... 324,244
E. Schrabetz................ 312,906
L. N. Nash.................. 312,499
                             331,078
                             331,079
                             331,080
                             331,210
D. S. Regan................. 320,285
S. Sintz.................... 315,082
G. M. Ward.................. 311,214

— 1886 —

C. H. Andrews—H. Williams... 341,538
G. C. Anthony............... 317,226
J. Atkinson................. 336,505
J. Charter.................. 335,564
J. H. Clark................. 347,469
G. Daimler (reissue)........ 10,750
E. Delamare—Deboutteville... 333,838

J. Hodgkinson—J. H. Dewhurst	347,603
E. J. J. Lenoir	345,596
J. J. E. Lenoir	335,462
P. Murray, Jr	351,393
	351,394
	351,395
L. H. Nash	334,039
	341,934
	341,935
N. E. Nash	340,435
J. F. Place	348,998
	348,999
N. B. Randall	355,101
A. L. Riker	349,858
C. Sintz	339,225
H. & C. E. Skinner	335,971
R. F. Smith	345,998
	347,656
J. Spiel	349,464
S. Wilcox	343,744
	343,745
L. H. Nash	334,038
	334,040
E. Korting	346,374
J. H. Clark	353,402
C. E. Skinner	352,368
	335,970
F. Bain	354,881
C. W. Baldwin	352,796
N. A. Otto	350,077
H. Robinson	346,687
N. A. Otto	335,038
J. P. Holland	337,000
	335,629
A. K. Rider	349,983
G. Daimler	334,109
J. Spiel	349,369
G. Ragot & G. Smyers	350,769
L. H. Nash	334,041

— 1887 —

J. Atkinson	367,496
C. W. Baldwin	368,444
	368,445
H. Campbell	367,184
J. Charter	356,447
	370,242
L. T. Cornell	359,920
F. W. Crossley	370,322
C. J. B. Gaume	374,056
F. W. Ofeldt	356,419
A Schmid—J. C. Beckfield	362,187
	371,793
Reissue	10,878

R. Van Kalkreuth	358,134
J. S. Wood	363,497
N. C. Bassett	359,552
T. Shaw	367,936
W. Gavillet—L. Martaresche	357,193
E. Korting	366,116
F. Von Martini	358,796
T. Backeljan	364,205
H. P. Holt—F. W. Crossley	370,258
N. A. Otto	365,701
F. W. Crossley—H. P. Holt—F. H. Anderson	363,508
B. F. Kadel	374,968

— 1888 —

H. T. Dawson	392,191
E. Delamare — Deboutteville, Reissue	10,951
H. Hartig	391,528
L. N. Hopkins	379,397
E. Korting	377,623
L. H. Nash	386,208
	386,210
	386,211
J. Noble	379,807
H. K. Shanck	376,212
	390,710
W. S. Sharpneck	391,486
C. Sintz	383,775
H. Skinner	389,608
R. F. Smith	377,962
G. W. Stewart	381,488
J. Bradley	386,233
J. R. Daly	392,109
L. H. Nash	386,214
	386,216
	386,212
	386,213
	386,215
	286,209
R. Bocklen	384,673
N. A. Otto	388,372
H. Williams	386,949
N. A. Otto	386,929
A. Rollason	391,338
	394,299
C. L. Seabury	393,080

— 1889 —

A. Schmid—J. C. Beckfield	403,294
J. J. R. Humes	400,850
C. W. Baldwin	407,320
	407,321

C. W. Baldwin	408,623
T. B. Barker	400,163
J. C. Beckfield	396,022
L. T. Cornell	406,263
W. E. Crist	417,471
H. J. Hartig	415,197
A. Histon	400,458
S. Lawson	399,907 / 399,908 / 402,749 / 402,750 / 402,751
J. Mathies	411,668
L. H. Nash	401,453 / 418,417
D. S. Regan	408,356
N. Rogers—J. A. Wharry	403,379
A. Schmid	396,238
C. Sintz	416,649
H. Tenting	402,363
W. von Oechelhaenser	417,759
C. White—A. R. Middleton	406,807
L. F. McNett	407,961
N. Rogers—J. A. Wharry	403,378
W. E. Crist	417,472
L. H. Nash	418,419
E. D. Deboutteville—L. P. C. Malandin	400,754 / 411,644
E. Capitaine	408,460
E. Korting	417,924
N. Rogers—J. A. Wharry	403,377 / 403,376
L. H. Nash	401,452
L. C. & B. Parker	401,204
E. Capitaine	408,459
I. F. Allman	411,211
N. Rogers—J. A. Wharry	403,380
J. C. Beckfield	417,624
S. Griffin	412,883
H. Hoelljies	408,483
L. H. Nash	418,418
E. Capitaine	406,160
C. S. A. H. Wiedling	398,108
J. J. Purnell	408,137
S. Wilcox	402,549
L. C. & B. Parker	403,367 / 405,795
W. J. Crossley	406,706
G. Daimler	418,112
J. Charter	415,446
N. A. Otto	407,234
M. V. Schiltz	399,569
A. Allmann—F. Küppermann	412,228
K. Gramm	415,908

— 1890 —

G. B. Brayton	432,260
W. D. & S. Priestman	430,038
E. Butler	423,214 / 437,973
H. Lindley & T. Browett	440,485
N. A. Otto	433,806 / 433,807
H. Campbell	428,801
G. McGee	432,638
J. Taylor	443,082
N. A. Otto	433,809 / 433,810 / 433,811 / 433,812 / 433,813 / 433,814 / 437,508 / 424,345
M. M. Barrett—J. F. Daly	434,695 / 430,504
F. Dürr	442,248
H. J. Baker	421,473
C. W. Baldwin	434,171
M. M. Barrett—J. F. Daly	430,505 / 430,506
J. C. Beckfield	432,720
J. C. Beckfield—A. Schmid	421,474 / 421,475 / 421,477
E. H. Gaze	437,776
J. Mohs	426,297
E. Quack	441,582
D. S. Regan (reissue)	11,068
A. Schmid—J. C. Beckfield	421,524
H. K. Shank	439,200
W. S. Sharpneck	441,028
C. Sintz	426,337
J. D. Smith	418,821
E. A. Sperry	433,551
J. R. Valentine—A. T. Grigg	425,116
C. W. Weiss	419,805 / 419,806
C. White—A. R. Middleton	438,209
J. J. Pearson—J. Kunze	428,858
G. H. Chappell (Rotary)	441,865
J. H. Eichler (Rotary)	442,963
G. E. Hibbard (Rotary)	424,000
W. S. Sharpneck (Rotary)	428,762
W. C. Rossney	420,169
E. F. Roberts	424,027
C. W. Baldwin	439,232
J. J. Pearson	426,736
J. W. Eisenhuth	436,936

PATENTS.

J. W. Eisenhuth	430,310
	430,312
G. B. Brayton	432,114
A. W. Schleicher—P. A. N. Winand	434,609
P. A. N. Winand—L. V. Goebbels	435,637
J. Roots	425,909
H. A. Stuart	439,702
N. A. Otto	437,507
C. von Lüde	435,439

— 1891 —

A. Harding	452,520
I. F. Allman	453,071
J. Charter	455,388
B. H. Coffee	446,851
P. T. Coffield—C. H. Poxson	456,284
E. W. Evans	452,568
J. Fielding	450,406
M. A. Graham	445,110
O. Kosztovits	448,924
G. W. Lewis	451,621
E. Narjot	448,989
B. C. Vanduzen	448,597
G. J. Weber	449,507
	444,031
M. M. Barrett	452,174
M. M. Barrett—J. F. Daly	463,435
D. D. & J. T. Hobbs	460,070
F. W. Lanchester	459,403
	459,404
	459,405
	465,480
L. G. Wolley	450,091
J. S. Connelly	457,459
	457,460
M. Loutsky	460,241
P. Neil—A. Janiot	462,447
H. Williams	457,020
B. C. Vanduzen	448,386
P. C. Sainsevain	461,802
G. Roberts	446,016
F. S. Durand	455,483
H. Schumm	458,073
H. Lindley	450,771
E. Kaselowsky	463,231
G. W. Lewis	451,620
A. Rollason—J. H. Hamilton	456,505
	456,853
	457,332
O. Lindner	453,446
L. Kessler	451,824
D. S. Regan	448,369

— 1892 —

J. Joyce	480,019
B. Stein	478,651
D. Best	484,727
J. Charter	472,106
J. A. Charter	473,293
	477,295
H. T. Dawson	466,331
E. W. Evans	488,165
J. W. Raymond	488,483
H. Warden	486,143
J. Wehrschmidt	484,168
C. W. Weiss	473,685
S. Withers—D. S. Covert	487,313
H. Schumm	488,093
E. I. Nichols	480,272
H. Schumm	482,202
A. Niemezyk	480,737
G. W. Weatherhogg	480,535

— 1893 —

F. E. Tremper	495,281
	503,016
J. S. Bigger	491,403
F. Cordenons	500,754
J. Foos—C. F. Endter	494,134
C. J. B. Gaume	501,881
W. W. Grant	497,239
C. F. Hirsch—A. Schilling	507,436
D. D. Hobbs	506,817
G. E. Hoyt	502,255
	510,140
S. Lawson	498,476
G. W. Lewis	511,535
W. von Oechelhaeuser — H. Junkers	508,833
C. W. Pinkney	499,935
	504,614
	505,327
	511,158
J. W. Raymond	491,855
C. Sintz	509,255
C. V. Walls	498,700
H. A. Weeks—G. W. Lewis	511,478
W. H. Worth	504,260
H. W. Tuttle	510,213
D. Best	496,718
C. W. Weiss	492,126
A. Niemezyk	508,042
C. B. Wattles	509,981
E. Delamare—Deboutteville—L. Melandin	511,593

18

H. Schumm.................... {497,689 / 510,712}
C. Stein...................... 511,661
P. H. Irgens.................. 505,767
H. Williams................... 490,006
B. Chatterton................. 505,751
A. Gray 504,723
W. Seck....................... 509,830

— 1894 —

J. Low—J. W. Gow............. 515,297
P. A. N. Winand.............. 525,828
A. J. Painter................. 523,369
W. S. Elliott, Jr............. 523,628
H. F. Frazer.................. 526,348
J. B. Carse................... {518,177 / 518,178}
B. H. Coffey.................. 514,211
H. T. Dawson.................. {513,486 / 530,508}
W. W. Grant................... 525,651
J. W. Hartley—J. Kerr......... 515,770
C. F. Hirsch.................. 526,837
F. Hirsch..................... {522,712 / 530,523}
C. S. Hisey................... 514,713
J. Labataille—J. J. Graff..... 517,821
D. C. Luce.................... 519,863
J. McGeorge................... 525,857
F. S. Mead.................... 528,006
H. B. Migliavacca............. 528,105
E. Narjot..................... 515,530
F. C. Olin.................... 525,358
J. & W. Paterson.............. 528,489
T. H. & J. T. H. Paul......... 530,237
H. Pokony..................... 514,271
S. D. Shepperd................ 521,443
H. Swain...................... 519,880
R. Thayer..................... 517,077
H. Voll....................... 527,635
J. Walrath.................... 522,811
F. Hirsch..................... 518,717
W. W. Grant................... 514,359
K. A. Jacobson................ 514,996
M. Lorois..................... 529,452
W. A. Shaw.................... 523,734
W. F. West.................... 513,289
W. Seck....................... 517,890
H. M. L. Crouan............... 515,116
H. H. Andrew—A. R. Bellamy {526,369 / 528,063}
H. Schumm..................... 528,115
H. Campbell................... 523,511

L. Crebessac.................. 530,161
R. B. Hain.................... 531,182

— 1895 —

G. W. Waltenbough............. 543,116
H. Schumm..................... 548,142
F. M. Underwood............... 542,743
F. S. Mead.................... 546,238
H. Thau....................... 545,553
A. J. Signor.................. 538,132
C. L. Ives.................... 534,886
M. L. Mery.................... 543,157
C. W. Weiss................... {543,163 / 543,165}
J. J. Norman.................. 548,922
J. J. Bordman................. 547,414
J. Bryan...................... 542,972
E. E. Butler.................. 546,110
J. A. Charter................. 532,314
F. W. C. Cock................. 544,210
F. W. Coen.................... 551,579
G. F. Conner.................. 548,628
F. E. Covey—G. W. Haines..... 532,869
W. L. Crouch—E. E. Pierce.... 535,815
J. Day........................ {543,614 / 544,214}
H. J. Dykes................... 539,122
J. Froelich................... 550,266
E. R. Gill.................... 536,029
H. H. Hennegin................ 545,502
F. Hirsch..................... 532,555
A. R. Holmes.................. 540,490
L. M. Johnston................ 538,680
J. W. Lambert................. {534,163 / 550,832}
H. A. Lauson—J. J. Norman—
  A. D. Nott.................. 550,451
F. S. Mead.................... {541,773 / 545,709}
F. P. Miller.................. 532,980
C. M. Rhodes.................. {531,861 / 540,923}
F. A. Rider—S. Vivian........ 533,922
B. L. Rinehart—B. M. Turner.. 552,332
C. Sintz...................... 539,710
E. J. Stoddard................ 533,754
H. Swain 535,964
G. Van Zandt.................. 537,253
C. V. Walls................... 537,370
G. J. Weber................... 534,354
H. A. Weeks................... 543,818
C. J. Weinman—E. E. Euchenhofer.................. 537,512

C. White—A. R. Middleton	545,995	C. Wagerell—A. A. Williams	555,355
D. Best	544,879	W. W. Grant	553,460
F. Burger	549,626		553,488
J. R. Bridges	548,772	S. M. Miller	553,352
J. W. Lambert	536,287	F. M. Underwood	553,181
G. W. Roth	552,263	W. D. & S. Priestman	552,718
W. R. Campbell	550,742	J. S. F. & E. Carter	552,686
B. W. Grist	545,125	L. J. Monahan—J. D. Termant	561,123
J. Robison	532,098	P. A. N. Winand	561,302
P. Burt—G. McGhee	550,674	H. L. Parker	560,920
G. W. Roth	539,923	J. W. Eisenhuth	558,369
F. S. Mead	544,586	G. Alderson	560,016
J. E. Weyman — A. J. & J. A. Drake	542,124	A. F. Rober	560,149
		L. H. Nash	563,051
P. Bilbault	532,412	T. M. Spaulding	562,673
A. R. Bellamy	536,997	L. S. Gardner	562,720
	537,963		558,943
O. Colborne	550,675	E. Kasalowsky	559,290
J. Robison	532,099	I. F. Allman	556,237
C. & A. Spiel	532,219	H. C. Baker	563,249
J. E. Friend	550,785	F. S. Mead	563,670
S. Griffin	542,410	A. W. Bodell	563,548
W. Seck	549,939	P. A. N. Winand	563,535
H. F. Wallmann	548,824	L. F. Allman	563,541
W. E. Gibbon	547,606	L. M. Burgeois, Jr	564,182
	535,914	A. J. Pierce	564,643
V. List—J. Kossakoff	536,090	E. N. Dickerson	564,684
	550,185		565,157
A. W. Brown	532,865	H. Swain	564,769
F. Mayer	549,677	J. Robison	565,033
F. W. Ofeldt	538,694	R. E. Olds—M. F. Bates	565,786
	540,757	B. Wolf	566,263
W. Lorenz	535,837	A. Barker	566,125
J. Robison	532,097	H. Ebbs	566,300
	532,100	G. H. Willets	567,530
		H. A. Winter	567,432
— 1896 —		H. Van Hoevenburgh	567,928
J. F. Duryea	557,469	C. D. Anderson	567,954
J. F. Daly & W. L. Corson	557,493	J. S. Klein	568,115
G. E. Hoyt	561,890	J. S. — R. D. — W. D. & C. H. Cundall	568,017
A. A. Hamerschlage	561,886	G. A. Thode	568,814
G. F. Eggerdinger and G. R. Swaine	561,774	F. C. Olin	569,386
G. W. Lamos	562,307		569,564
Fred Mex	562,230	H. A. Winter	569,530
H. G. Carnell	533,662	C. J. Weinman—E. E. Euchenhofer	569,365
	556,086	H. Schumm	569,942
F. W. Mellars	556,195	H. C. Hart	569,918
C. J. Weinman—E.E. Euchenhoffer	555,717	M. W. Weir	569,694
	555,791	T. von Querfurth	569,672
F. W. Crossley & J. Atkinson	555,898	R. E. Olds	570,263
M. G. Nixon	559,399	E. J. Pennington	570,440
J. M. Worth	559,017		570,441
G. L. Thomas	558,749		

R. Rolfson	570,649	E. E. Ludi	572,209
L. Gathman	570,470	E. Capitaine	572,498
E. Prouty	570,500	F. J. Rettig	573,296
C. W. Pinkney	571,239	F. E. Culver	573,209
C. A. Kunzel, Jr	571,447	S. M. Balzer	573,174
G. W. Lewis	571,534	J. Charter, Jr	573,762
F. C. Olin	571,495	G. S. Tiffany	573,628
E. Rappe	571,498	M. F. Underwood	574,183
M. Blakey	571,966	J. W. Eisenhuth	574,311
J. F. Duryea	572,051		

INDEX.

A

ABSOLUTE zero, 9

B

BATTERIES, 146
Beau de Rocha, 3, 4
Boyle's law, 8
Brake, Prony, 90
    rope, 93
    strap, 92
Brown's gas-vacuum engine, 3
Bunsen burner, 65

C

CARD, Atkinson, 28, 137
    combustion, 109
    full-load, 29
    half-load, 30
    typical, 31
    variable, 25
Carburetters, circular, 47
    Daimler, 50
    Gilbert & Barker, 52
    ventilating, 48
Causes of loss in motors, 33
Clerk, Dugald, experiments, 22
Coefficient of expansion, 10
Combustion rate, 29
    retarded, 25
Comparisons, Crossley, etc., 36-38
Cost of operation, 120
Cylinder, capacity, 54
    dimensions, 1, 55

E

ECONOMY, conditions, 27
    electric lighting, 37-39

Efficiencies, 13, 19, 20, 22, 23, 24, 26, 102
Electric generator, 77
    lighting economy, 37-39
Electrodes, 148
Explosive mixtures, 14, 16, 26
    mixture diagram, 15

F

FORMULAS for pressures, 11
    for temperature, 11
    for volume, 11

G

GAS ENGINE,
    the Allman, 181
    the American motor, 203
    the American, 170
    the Atkinson, 135
    the Backus, 175
    the Bollee, 264
    the Charter, 120
    the Climax, 244
    the Daimler, 208
    the Dayton, 149
    the Duryea, 258
    the Economic, 112
    the Facile, 249
    the Fairbanks-Morse, 159
    the Foos, 146
    the gasoline bicycle, 259
    the Hartig, 178
    the Hicks, 200
    the Hornsby-Akroyd, 240
    the Hydrocarbon Motor, 254
    the Lambert, 195
    the Lawson, 234
    the Nash, 184

the New Era, 114
the New York Motor, 246
the Olds, 219
the Pierce, 117
the Priestman, 228
the Prouty, 193
the Racine, 238
the Raymond, 128
the Ruger, 169
the Simplex, 251
the Sintz, 132
the Springfield, 141
the Star, 206
the Victor, 150
the Vreeland, 174
the Weber, 222
the Webster, 139
the Wing, 265
the White & Middleton, 253
the Wolverine, 154
Gas, natural, 43
 producer, 44
 semi-water, 45
 water, 45
Gasoline regulator, 144
 evaporation, 53
 pump, 117
Gay Lussac's law, 9
Globular cylinder head, 34
Governors,
 inertia, 61
 New Era, 116
 pendulum, 63
 pick-blade, 60
 Robey, 58
 vibrating, 62

H

Heat efficiencies, 102
 of combustion, 12
 of compression, 105
 units, 35, 42
 value of gas, 21
Historical, 3

I

Ideal card, 23
Ignition timing-valves, 78, 79

Igniters electric, 72, 77, 145, 148
 flame, 65, 66, 68
 hot-tube, 58, 71
Indicator, 96
 cards, 19, 25, 31, 109, 137
Introduction, 1

J

Joule's law, 12

L

Lenoir indicator card, 19
 motor, 3
Light in cylinder, 8
Lubricators, 81, 82

M

Management of explosive motors,
Material of power, 41
Measurement of power, 89
 of speed, 94
Mechanical equivalent, 9
Mufflers, 56

O

Otto, 3, 4
 card, 27

P

Patents, number of, 4, 5
 list of, 269
Petroleum distillate, 46
 products, 45
Piston speed, 26
Prony brake, 90

R

Ratios, volume, pressure, heat,
 of expansion, 10, 11
Reducing-pulley, 99
Rope brake, 93

S

Shrinkage of charge, 30
Sparking coil, 75

Specific heat, gas, and air, 20
Speed measurement, 94
Spherical combustion chamber, 34
Strap brake, 92
Stratification, 13

T

TACHOMETER, 95
Temperature of cooling water, 34
    of combustion, 108
Testing explosive engines, 107
Timing-valves, 78, 79
Theory of gas and gasoline engines, 7

U

UTILIZATION of heat in gas engines, 18

Useful effect, 26

V

VALVE chest, 116
Vaporizer, 49
Velocity of explosion, 14
Vibration floors and buildings, 100
Volumes, piston, 106

W

WALL cooling, 25
    surface, 28
Water tank, 35
Worm gear, 116

Z

ZERO, absolute, 9

www.ingramcontent.com/pod-product-compliance
Lightning Source LLC
Chambersburg PA
CBHW032116230426
43672CB00009B/1754